USING SMART CARDS TO GAIN MARKET SHARE

❖

USING SMART CARDS TO GAIN MARKET SHARE

❖

Aneace Haddad

Published by
Gower Publishing Limited
Gower House
Croft Road
Aldershot
Hampshire GU11 3HR
England

Gower
Old Post Road
Brookfield
Vermont 05036
USA

British Library Cataloguing in Publication Data
Haddad, Aneace
Using smart cards to gain market share
1. Smart cards 2. Market share 3. Marketing – Management
4. Sales promotion
I. Title
658.8'2

ISBN 0 566 08315 9

Library of Congress Cataloging-in-Publication Data
Haddad, Aneace.
Using smart cards to gain market share / Aneace Haddad.
p. cm.
ISBN 0-566-08315-9 (hc.)
1. Customer services. 2. Smart cards. 3. Marketing. I. Title.

HF5415.5 .H24 2000
658.8'83—dc21 99-057512

Typeset in Garamond Light by Bournemouth Colour Press, Parkstone and printed in Great Britain by MPG Books Ltd, Bodmin.

CONTENTS

❖

LIST OF FIGURES AND TABLES

❖

Figures

Tables

ACKNOWLEDGEMENTS

❖

This book would not have been possible without the extraordinary intellectual support provided by my colleagues at Welcome Real-time. Bernard Chevalier has been at my side for many years as we worked through the details of this exciting new industry. Jean-Yves Grall's background with technology brands proved to be a major catalyst and a fundamental foundation for powerful new differentiation strategies. Pierre-Philippe Cormeraie commented on early drafts of the manuscript and provided insight into how to synthesise what would otherwise be a very complex business case. I would also like to personally thank Thierry Sert, Nicolas Lefort and the rest of the R&D team members who have painstakingly brought to life Welcome Real-time's vision.

Our clients deserve my utmost gratitude and appreciation for their leadership and pioneering efforts. They know who they are.

Frédéric Chevalier was an early believer and has provided valuable support, creative insight and encouragement throughout the years. Thank you.

Last but not least, this book is the result of many years of hard earned practical experience that would have been impossible to acquire without the constant support of my wife, Sophie.

AH

1

GAINING MARKET SHARE AT THE MOMENT OF PURCHASE

❖

No longer are our competition American Express and MasterCard. The real competitor is cash and checks.

Carl Pascarella, CEO, Visa USA[1]

Every time you use your credit card a complex interaction takes place between yourself, the retailer, your respective banks and a number of intermediaries that allows the process to function smoothly.

The age-old exchange between buyer and seller, culminating in the transfer of money from one to the other, has gone through a long evolutionary process to arrive at today's complex system.

Taking the long view, the moment of purchase has been under constant evolution ever since people began exchanging valuables among each other. Non-barter payment methods were invented thousands of years ago. Long before that, people were already negotiating their prices based on volume purchasing. You buy more of my goats today, I will sell them to you at a lower price per goat. A bird in the hand is worth two in the bush. The pragmatic economics of customer relationship marketing and loyalty programs goes back to the dawn of human civilization.

Information technology is the latest element to enter the evolutionary process at the moment of purchase. This began over 25 years ago. Now, smart cards, as a powerful agent of information technology, have entered the scene and promise to greatly accelerate the transformation of even the substance of money.

Once information technology was introduced into the buyer/seller interaction, the evolutionary process accelerated rapidly. An explosion of techniques has recently allowed the retailer to better manage the selling process through point-of-sale terminals, barcode scanners and on-line just-in-time inventory management. Banks brought significant improvement to the payment process by placing magnetic strips on the backs of credit cards and installing terminals that could automatically request credit approvals and record the transactions. This was all accomplished over a twenty-five year period, a nanosecond in the history of buyer/seller transactions.

Recent innovations suggest that this is only the beginning. The industry is continuously moving forward, accelerating relentlessly to further integrate computers and communications technologies, coming together seamlessly at the moment of purchase, an event that often lasts only a few seconds. Smart cards are entering into this complex moment, promising further improvements in payment processing – improvements like better fraud control, faster transactions, reduced processing costs. What is even more exciting is that smart cards also promise to deliver completely new benefits that will improve the relationship between retailers and their customers. All in that short yet tremendously important moment when the buyer and seller interact.

The moment of purchase is being transformed by the convergence of payment methods, smart cards and customer relationship marketing. It is certainly becoming increasingly complex behind the scenes, but it is also becoming simpler for cashiers and customers. And the moment of purchase is happening faster than ever before. Punch in an amount, insert the card, type a PIN code, and in several seconds the transaction is over, while behind the scenes hundreds of complex operations were performed and thousands of others were triggered.

The most profound technologies are those that disappear, because paradoxically they have achieved a level of complexity that allows them to disappear. Antoine de Saint Exupery once wrote, 'Perfection is achieved, not when there is nothing left to add, but when there is nothing left to take away'. Automobiles at the start of this century were exclusively for mechanical engineers and tinkerers, since each model had its own specific quirks and the technology had to be constantly supervised and tweaked in order to continue functioning. Today's cars are several orders of magnitude more complex, yet they are so simple to drive that most people have no idea what type of engine is under their bonnet. Take the ubiquitous technology of alphabets. Writing was originally an élite capability, reserved for government bureaucrats. Now, we are no longer aware of performing what was once considered to be a highly complex and almost magical feat as we constantly read words everywhere, even pasted in the form of brands on goods we see and use every day. All of our tools have progressively evolved in this direction ever since we began making them.

Increasing complexity, power and efficiency becomes thoroughly irresistible when combined with significant improvements in ease of use. This is what is happening with the convergence of payment methods, smart cards and customer relationship marketing. Each of these has been on separate trajectories that lead to their current meeting point at the moment of purchase.

Evolutions in payment methods

Eight thousand years ago, as hunter gatherer tribes began settling down in a farming lifestyle, they were already familiar with the only payment method available. Money was whatever they grew, hunted, or made. Nearly three thousand years ago the first coins appeared, which was a great leap forward in terms of portability and convenience. Three hundred years ago printed paper money appeared. Soon after that, cheques were invented. The newly discovered ability to move money readily and access it without carrying it paved the way for the use of plastic cards for payment.

In the 1940s, customers could not easily buy expensive items without going to their bank and filling out complicated loan applications. You would walk into a department store, select a new refrigerator, then have to leave the store and visit your bank. Along the way, you might notice another department store and end up buying your refrigerator there. You might even simply change your mind and not buy anything at all. The system was frustrating to the customer, as well as to the retailer (since there was a risk of losing your business) and even to the banker (who in many cases could have done more business if things were less complicated). The answer was the revolving credit card, which so successfully addressed the needs of the banker, the retailer and the customer that it launched an entire industry.

The payment card industry began in the 1940s and has successfully initiated a number of revolutions over the past 50 years, with the introduction of numerous innovations in payment card products. Common card acceptance brands, Visa and MasterCard, allowed customers to use their cards first nationally then internationally. A magnetic strip on the back of each card, containing an electronic copy of the information embossed on the card, allowed terminals to automatically call a host to obtain a credit authorization, so the retailer no longer had to call an operator. Debit cards were introduced to replace cheques and to allow a bank's customers a Visa card even if they did not want, or could not have, a standard credit card. Today, smart cards are just one of the most recent of a long line of successful payment card innovations. Financial institutions expect smart cards to help take further market share away from cash and cheques, reduce fraud and processing costs, and better compete against each other.

Here is a description of the type of payment product that MasterCard is aiming at:[2]

> Ancient forms of value exchange – livestock, giant stones, spices, gold – were anything but convenient. Currency represented a fundamental shift away from these early varieties of money: You had something that represented the wealth you owned. The development of checks and bank drafts made carrying cumbersome amounts of currency less necessary. Today, credit and debit cards make life more convenient; they're easy to carry and use, and enable you to keep track of your spending by providing monthly statements. Virtual – or electronic – money takes the convenience concept a step further: You'll be able to access your bank account from anywhere in the world, obtain special services and benefits, and make purchases safely.
>
> In the future, convenience will drive the form of payment people choose. Credit cards let you buy now and pay later, if that's your preference. Debit cards let you buy now and pay immediately through your checking account. Electronic cash lets you 'pay in advance' by loading value onto the card, then making purchases using the stored value. 'Smart cards' will offer even more conveniences that cash and checks can't. A single card will store information – health insurance data, motor vehicle permits, currency exchange rates, loyalty programs – and operate as virtual money.

Since the payment card industry has already successfully integrated numerous innovations over several decades, much data is available to better understand how the move to smart cards is likely to succeed. A surprisingly small number of intuitively obvious factors are required in order for a new financial card product to succeed.

A new card product can substantially increase its chances of success by simply addressing the core motivations of each of the main participants: banks want to increase their share of transactions, retailers want to increase their share of profitable customers and customers want access to high quality products and services at the lowest price. The most successful products satisfy all three requirements. This is just plain common sense. Unsuccessful products satisfy a single requirement, perhaps two.

In light of this straightforward analysis, one can easily understand the market's lack of interest in electronic cash or 'e-purse' schemes that simply attempt to satisfy the bank's motivation to capture a larger share of transactions without adequately addressing the needs of retailers and cardholders. We will show how smart cards can be used in such a way as to maximize the card's chances of success and allow the card issuer to enjoy lasting benefits.

The arrival of smart cards

Smart cards are credit cards with an embedded microprocessor chip. The microprocessor's memory module can hold many times the information contained in a credit card's magnetic strip. Data is also far more secure, since access to files can be protected by keys. Most smart card microprocessors come with built in encryption engines based on powerful algorithms.

The smart card was invented over twenty years ago in France by Roland Moreno, an engineer tinkering with ways to automate identification procedures. The plastic credit card format was not his first choice. He tried using rings at one point, but discovered that it was hard to insert a knuckle precisely in the reader and keep it still long enough for the transaction to take place. This is all part of the mythology that most people in the industry are familiar with. But what is the real story behind the success of smart cards in France?

New infrastructure technology requires the scrapping of existing equipment – cards, terminals, networks – and their replacement with new, expensive equipment. Whole industries must find justification to move forward or the budding technology simply dies quietly and disappears without anyone noticing. Encouraging the machine to begin inching forward, building momentum, is no small task for a single person.

In fact, the French government played a central role with smart cards by acting as both supplier and buyer of the new technology. Government-owned chip and computer manufacturer Bull was encouraged to manufacture cards and terminals, while France's telephone company became a big first customer for prepaid disposable telephone cards. This seeded the market and triggered the proliferation of French smart card and equipment companies.

At around the same time, French banks were faced with a dilemma. They were losing ground to large retailers who were launching their own credit cards and refusing to accept bank cards, as a way to cut out the 2–3 per cent processing fees. French banks were quickly forced to find a way of bringing fees down to a level acceptable to larger retailers. This became such a vital concern that all banks came together as a single industry coalition, whose primary objective was to protect the French banking industry's overall market share. Smart cards were chosen as a technology that can dramatically reduce fraud and transaction processing costs, the main causes of high fees. Pilots were launched, tweaked, and made to function correctly. Industry pricing was established across all players. Rules of competition were defined in order to ensure long-term cooperation among banks and avoid creating significantly different card products that might fracture the banking industry and again weaken it.

Of course, this solution would possibly be illegal elsewhere because the French banking community has built something that resembles a cartel. Nevertheless, the results are striking. Processing fees were slashed to 0.8 per cent, where they still stand today. One after another, large retailers began accepting bank cards. Many retailers even shelved their private card projects. The French banking industry was saved.

In France, 30 million Visa, MasterCard and Carte Bancaire debit cards are all equipped with microprocessor chips. Since smart cards became the general form of payment in 1992, bank card fraud has decreased by over 70 per cent. Transaction processing costs have also been reduced dramatically, because the smart card's off-line technology reduces the number of connections required between the terminal and the central processing system. The workload on the central system is lighter, telecommunication costs are lower, and the checkout counter in the store runs faster and more smoothly.

The smart card is also beginning to open a new market in transactions involving small sums of money which are now paid in cash. The cost of handling cash generally amounts to 2.6 per cent of the purchase amount, an annual cost of $60 billion in the United States alone. Electronic cash is a leading market that many financial institutions hope to play a part in, if not dominate.

These are some of the reasons why Visa and MasterCard have been working together worldwide to define common specifications for enabling banks to add a microprocessor to their payment cards.

According to Edmund P. Jensen, President and CEO of Visa International:

> Cash and checks will become endangered species. Demand for convenience, globalization and large new markets are driving technological advances to create new ways to pay. Chip cards will transform the way we shop, allowing consumers to conduct safe electronic transactions anytime and anywhere in the world.

The success of French banks has now prompted financial institutions in many countries to begin moving to smart cards. Live tests began outside France around 1995. Initially, many financial institutions were attracted to the electronic purse, or stored value systems that use bits in the card's microprocessor chip to represent cash. UK banks, led by Natwest, launched Mondex and immediately created an international marketing phenomenon around that e-purse, despite lack of interest by UK consumers and retailers. MasterCard purchased 51 per cent of Mondex, which had a snowballing effect on the number of pilots being carried out worldwide. Visa Cash has also been piloted in a number of different environments, one of the earliest being the 1996 Olympics in Atlanta where NationsBank, now Bank of America, First Union and Wachovia issued thousands of smart cards.

Attendees used the cards to buy soft drinks, sandwiches and other small items. As of yet, the most successful e-purse product by far is Proton, which was created by a Belgian bank consortium. In 1999 Proton had become by and large the world's foremost e-purse with 30 million cards distributed by over 250 banks, and accepted by 200,000 terminals in 15 countries. Proton is one of the few e-purse systems to have been used nationwide in Belgium, Holland, Switzerland and Sweden.

The e-purse is not the only smart card application being tested. A number of financial institutions in Asia, such as Standard Chartered Bank in Hong Kong, have been focusing on using smart cards to add value to their core credit card business. Several pilots have been launched with credit and debit cards linked to other applications, primarily customer loyalty.

In the UK, a massive program to replace conventional credit cards with smart cards began in April 1999, in an attempt by UK banks and financial institutions to eliminate fraud. Credit card fraud in the UK has been growing annually by 20 per cent and is expected to cost businesses as much as £300 million per year by 2002.[3] In an effort to combat this, UK card issuers will convert an estimated 100 million credit and debit cards to chip over a five-year period. Given the success of the French banking industry's earlier move to smart cards, UK bankers expect to cut credit card fraud by as much as half.

Every single smart card pilot has been valuable for the industry because each has revealed a wealth of information on what customers and retailers want in a new generation card product. Customers want something simple, convenient, easy to use and very easy to understand. Retailers want the same thing, and more. They especially want to know that the new payment method they are being asked to accept will allow them to increase their revenues. This discovery has prompted card issuers worldwide to scramble to offer loyalty solutions that help retailers recognize and reward loyal customers. New card products that don't adequately address these requirements simply cannot build enough steam to achieve critical mass.

Early smart card pilots were all single-function implementations, primarily electronic purse. The industry has matured in a few very short years. Extensive communication and collaboration among all participants has helped pioneers learn from each other and arrive quickly at an industry-wide consensus. Today, the vast majority of financial institutions are now considering multifunction implementations combining a traditional payment mechanism like credit or debit along with a customer loyalty function and perhaps an electronic purse.

Understanding of the market has matured. The technology has been proven. The business case for credit cards is of course already well understood, but so is the business case for customer loyalty, since virtually all banks have been more or less involved in co-branded and private card programs for retailers for close to a decade. Mature software applications exist for the primary functions card issuers want. Existing off-the-shelf card technology already supports multiple functions on the same card, at a low cost. The context is right, the time is right, the technology is right and the players are in place. All of these elements work together to begin nudging the industry toward mainstream adoption.

The bank card associations, Visa and MasterCard, are becoming more aggressive in promoting smart cards. More and more individual card issuers are approaching the launch of very large pilots of several million cards. Banks in some regions are even on their way to major national deployment. Requests for proposals are accelerating, with specifications for multimillion card implementations that were unheard of just one or two years ago. In terms at least of interest and commitment to moving forward, critical mass is building among financial institutions. Now, a prediction. The first to succeed in building critical mass among retailers and consumers will quickly discover that they have succeeded in carving out a comfortable leadership position that will prove hard to topple for many years to come.

Customer relationship marketing

In the 1980s, first in the US then over much of western Europe, new retail locations became increasingly hard to come by. For decades prior to that, one of the easiest and most effective ways for a retailer to increase sales and profits had been simply to open new stores in new locations. Once that became less of an option, they had to find new ways to increase revenue. One solution was to give customers an explicit incentive to be loyal to the retailer's store as opposed to the competitor's store. This has come to be known as 'customer relationship marketing' or, more popularly, loyalty.

A well defined and well managed loyalty card program, launched before competitors join in the game, often proves to be a formidable weapon that can deliver spectacular results. In just a few years, UK supermarket retailer Tesco was able to dethrone Sainsbury's, a company with a long and famous history that was the undisputed market leader at the beginning of the 1990s. Tesco proved adept at introducing a range of innovative new shopping services to entice customers. It launched a highly effective marketing campaign centred around the Clubcard, Tesco's loyalty card. Lord Sainsbury famously dismissed the idea as nothing more than an 'electronic version of Greenshield stamps'. He lived to regret that comment. Sainsbury's was forced to make an embarrassing turnaround and introduce its own loyalty card when the Clubcard became a roaring success. But it was already too late. Sainsbury's has been forced to play catch-up, with decidedly mixed results. In the same week that Sainsbury's announced layoffs at its head office, Tesco announced 20 000 new jobs and an ambitious overseas expansion plan.[4]

Loyalty and frequent buyer programs are mature applications that have already been linked to hundreds of millions of magnetic strip payment cards since the 1940s, when large retailers like Sears and JC Penney's in the US began offering charge cards to their customers. Then, in the 1980s, US banks and financial institutions began tying their credit cards to airline loyalty programs. Single-partner or co-branded programs let cardholders earn travel on one airline only. Marine Midland was first in 1986, followed by Citibank/American Airlines and First Chicago/United Airlines in 1987. Multi-partner programs let cardholders earn travel on multiple, selected airlines. Diners Club was first to launch such a program when it introduced 'Club Rewards' in 1984. American Express followed with 'Membership Rewards' in 1991, while Bank of Montreal and Citibank launched 'Air Miles' in 1992.

Loyalty programs have generally proved to be an effective way to differentiate card products and compete on benefits rather than interest rates and transaction fees. Once banks began offering airline rewards, the whole industry quickly saw the benefits of such programs and companies scrambled to establish alliances with leading airlines. Opportunities for co-brand relationships in all retail sectors, including airlines, are obviously limited. Ten years after the launch of the first co-branded airline card, there were virtually no more exclusive agreements available anywhere in the world. Banks and financial institutions had completely secured the market.

In the same way that credit cards proved to be a great success because of their ability to simultaneously satisfy the needs of customers, retailers and bankers, the emergence of customer loyalty programs also proved successful because of their similar ability to address distinct market motivations.

There are three main types of loyalty programs: private cards, co-branded cards and punch cards. Private cards were the first to be addressed by banks and financial institutions who were eager to offer credit-processing services to very large, centrally managed retailers. This proved to be a big market. Once the private card market peaked and it became difficult to find new business, financial institutions then turned to co-branded cards for new revenue streams, which developed into an even larger market. Now it too has peaked, and financial institutions have just begun setting their sights on the largest of all categories of loyalty programs, punch cards, which have yet to be linked to payment cards. This new market will be even bigger and more mainstream than private and co-branded cards.

The private card market is big, but it is also mature

A few card companies seeking new markets for their payment card services first began addressing potential private card issuers. At the time, many leading credit card issuers refused to address retailers' private card requirements, since by definition the bank's brand does not appear on the card. Retailers wanted access to financial services only, an approach that many banks feared. Nevertheless, a few financial institutions, like GE Capital and others, went ahead and offered financial services to retailers and quickly established what turned out to be a large market. At the end of 1997, third-party processors managed 150 million private credit cards for US retailers.

Private cards are issued by large retailers who are able to justify a customer keeping their card in their wallet. Transactions tend to be of a higher value and represent a higher share of a customer's overall spending. Virtually all large, centralized retailers have issued private cards at one time or another, quite often as a reaction to a competitor's program, which is why many of these cards are essentially identical. Although a good private card program can be very attractive to customers, few retailers have such a distinctive brand that customers want to keep that retailer's private card in their wallet and use it regularly. The vast majority of retailers are unable to justify issuing private cards, simply because it is so hard to convince a customer to apply for the card, keep it in their wallet and use it on a regular basis. For this reason many of the larger retailers that chose not to create their own private card have become issuers of co-branded cards.

The market for co-branded cards is even bigger, but it too is very mature

Once the private card market was mature and the primary players were in place and difficult to compete with, financial institutions turned to co-branded cards for new growth opportunities, which proved to be an even larger market. In 1994, about 20 per cent of all Visa and MasterCard transactions were made with co-branded cards and by 1997 the number was estimated to have reached 50 per cent.

With a co-branded card, the retailer's brand is prominent on the front of the card, next to a general purpose payment brand like Visa, MasterCard or American Express. These cards are easier to deploy, maintain and keep active, because they can be used wherever a standard credit card is accepted. It is usually easier to convince a customer to keep a co-branded card in their wallet than it is to get them to hold on to a private card, because co-branded cards tend to be more useful to customers than private cards. For this reason there are a larger number of co-branded card programs than there are private programs.

Today, the co-branded market is also mature. The players are well established and it has become difficult to compete. The private card market has recently undergone extensive consolidation, with smaller players selling their card base to larger players. The same has also been happening with co-branded cards, but to a slightly lesser degree. Competition is cut-throat and margins are razor-thin.

Mega-mergers between banks are now being engineered in Europe and many other regions, following the corporate combinations that have been witnessed in the US. In many industries, consolidation follows commoditization, as was evident decades ago in electricity, the motor industry, oil, steel and railroads. More recently, we have seen commoditization and consolidation in telecommunications, banking, airlines, media and retail – industries whose products had been reduced to sheer volume economics. These industries value brand new products that do not react merely to volume economics, and that can help open whole new markets larger than existing markets.

Punch cards, the biggest loyalty program market, are next to be integrated with payment cards

The largest loyalty category by far, the most mainstream and the one that has not yet been linked to payment cards, is the simple punch card that lets a customer earn a free item or discount after a certain number of visits or amount of money spent.

How many frequent-customer punch cards do you have in your wallet right now? Your favourite snack bar might have given you one that offers a free sandwich after buying several. You may also have a card from your local coffee shop, your dry cleaners and perhaps the video rental shop. Cinemas may give you one for free popcorn or a drink. Countless fast food restaurants use the technique to offer a free meal. Car parks offer free parking after paying for a certain number of days.

Besides the fact that all of these retailers use loyalty cards that are punched or stamped at each visit, they all have another thing in common. Generally, they don't accept credit cards. They are primarily cash and cheque retailers.

A general rule that emerges is that private store cards and co-branded cards tend to be used by large, centralized retailers that enjoy higher transaction amounts and that already accept credit cards. Punch cards are used by smaller retailers and franchisees, retailers that have smaller transaction amounts and that don't currently accept credit cards. Many large, centralized retailers have already launched their own credit cards, which are successful when the retailer is able to justify wallet presence. Punch card sector retailers are generally unable to persuade customers to sign up for their own private card and keep the card in their wallets – other than die-hard customers who are already solidly attached to the retailer.

The differences between retailers that issue private or co-branded cards and those that issue punch cards are considerable. Financial institutions that want to increase the use of credit cards among smaller retailers need to address those differences head on. By doing so, they will discover very large new opportunities.

The electronic purse is not necessarily the only way to extend a card payment mechanism among cash and cheque retailers. In many cases, simply using a new generation credit card with a chip for off-line transactions and for managing electronic punch cards, will prove to be a powerful combination.

Smart payment cards offer issuers new opportunity for differentiation through their ability to integrate dozens of electronic punch cards. The smart card acts as a wallet with lots of slots in it, ready to receive electronic punch cards from many different retailers. Using off-the-shelf, low cost, smart credit card technology, each retailer's electronic punch card can be added automatically to a customer's smart card the first time that customer shops at that retailer's store. At every visit, the card is electronically punched. Once the customer has reached the number of visits or amount of money spent necessary to receive the free item or service, a coupon is automatically printed out at the payment terminal. Most smart payment cards available today can support several dozen electronic punch cards at the same time. No two customers will have the same set of punch cards, simply because no two customers shop at exactly the same set of stores. As customers change their buying habits and choose to shop at new retailers, the punch cards automatically adapt to the cardholder's behaviour: as punch cards expire, they are simply replaced with new punch cards. Existing technology is already capable of supporting very large numbers of customers and retailers without fear of the cards becoming overloaded and without requiring customers to do complicated maintenance work, like choosing which programs they want and which they want to delete.

Fast food restaurants, coffee shops and dry cleaners are not the only companies that use punch cards. The technique has been around for ages and has become very widespread. Consumer goods manufacturers have for decades dreamt of loyalty techniques that would allow them to give discounts to customers after a certain number of purchases. Generations of people have cut out proof of purchase seals from cereal boxes, coffee labels, detergent and countless other products and sent them back to the manufacturer in exchange for a coupon for a free item. The technique is complicated, messy and very expensive. The same core group of fringe consumers participate over and over again in these programs. These are major drawbacks. Smart cards can help bring these techniques into the mainstream. Why not allow Procter & Gamble to run its own electronic punch card program, exactly like a coffee shop's program? By running the program on the customer's smart credit card, administration instantly becomes very simple. The customer pays with the card, and the receipt simply shows an additional line of information that tells the customer how much more they need to spend with Procter & Gamble in order to receive their gift certificate or electronic coupon. A real need exists for this type of program. Brand manufacturers will open up large budgets for strategic programs like electronic punch cards that help them at last reward their loyal customers over time, as opposed to simply offering coupons through the Sunday newspaper.

The economics of loyalty

The economics surrounding loyalty punch cards are simple. In essence, the retailer rewards you for concentrating your purchases in their store. This is the same economics that volume retailing is based on. If you buy a big box of detergent the cost per ounce is less than what you would pay for a small box. A large tin of coffee generally costs less per ounce than a small jar of coffee. One hundred PCs purchased at the same time, or over a one-year period, cost less per PC than if you only bought ten, simply because there are fixed costs related to each transaction, regardless of the size of the transaction. High marketing costs are necessary to find a new customer, bring him into your store, convince him that your products and services are satisfactory and process the first order. Once that's done, all subsequent orders enjoy lower transaction costs and profitability is generated by reduced servicing costs, less price sensitivity, increased spending and favourable recommendations passed on to other potential buyers. On a per unit basis, volume transactions have lower fixed costs than smaller transactions.

Punch cards are based on age-old economics that reward volume purchasing by sharing those savings with the customer. All successful loyalty programs are based on these same simple economics. More complex concepts have been developed by marketing firms, building subtleties on top of the basic foundations of volume pricing. Lifetime value, for example, is the predicted future revenue expected to be generated by a customer based on that customer's current behaviour. To calculate lifetime value, a particular customer's prior purchases are summarized according to three sets of parameters that together quantify behaviour: how recently the customer last made a purchase, how frequently purchases were made in a given period, and what cumulative amount was spent over that same period. These parameters are often referred to as 'RFM' – recency, frequency and monetary value.

Lifetime value and RFM are certainly important concepts, but they should be seen for what they really are. Quite simply, they are different ways to measure volume purchasing. Loyalty marketing may be shrouded in jargon, but the basic economics surrounding loyalty are no more complicated than those surrounding the simplest volume pricing policies.

Direct marketing professionals use lifetime value and RFM as customer segmentation methods to target mailings. Using centralized databases and sophisticated data-mining software, customers are ranked according to how recently they last shopped, how many purchases they made over a given period and how much they spent. Coupons, gift certificates and rewards can then be sent to customers based on individual or group behaviour profiles. The method is well established and relatively effective for higher value purchases capable of justifying the high cost of operating the system.

Companies that practise loyalty marketing for lower value purchases often cut costs by avoiding expensive database marketing techniques. Ironically, many program operators neglect the volume pricing aspect of loyalty once they discover the high costs necessary to do it properly through direct mail. These programs do not treat loyal customers better, since they cannot recognize one customer from another. This obviously defeats the purpose of running a loyalty program. Lack of differentiation has resulted in some of the failed programs that have been roundly criticized by customers and the press. Smart cards can significantly reduce direct marketing costs, thereby paving the way for successful application of loyalty marketing techniques to much larger market segments and product categories.

Airline frequent flier programs appear to be quite different from the simple punch card. At first glance. A closer look reveals precisely the same volume purchasing economics that exploit diminishing fixed costs per transaction. Repeat customers are less expensive because acquisition costs are spread out over more transactions. Repeat customers also tend to buy more seats at full price and more higher margin products, which is why in the airline industry, as in most other retail sectors, no matter how different they might be, the top 30 per cent of customers represent 75 per cent of sales while the bottom 30 per cent represent 3 per cent of sales.

Volume purchasing in virtually all industries and markets displays the same characteristics. Higher volume purchases cost proportionally less in fixed overhead and represent a proportionally larger share of profits.

Loyalty is nothing more mysterious than this. It is simply a way for customers to enter into a business agreement with a supplier who essentially says, 'If you spend more over time with me and buy more product more often, you pay less'. In business-to-business relationships, a vendor that does not offer volume pricing would be considered strange. In business-to-business this is not called customer loyalty, but I wouldn't know what else to call it.

Many US supermarkets have two prices next to products, the normal price, and the special price you pay if you spend at least $20 today. Analysis of the details of each transaction shows that larger shopping baskets represent a proportionally higher share of margins. Discount supermarkets were originally designed for customers who shop around once a week, whose transactions were naturally larger, usually well above the $20 advertised today. As more and more supermarkets were built, bringing them closer and closer to people's homes, and more and more began staying open 7 days a week and even 24 hours a day, customers began treating supermarkets as convenience stores. Run down to the store for a pint of milk. Stop by the supermarket on the way home from work for a pack of beer and maybe a video. When customers shop at traditional convenience stores like 7 Eleven, they are accustomed to paying significantly higher prices than in traditional supermarkets. This has prompted many grocers to say, 'If customers appreciate our stores as convenience stores, they should be willing to pay convenience store prices'.

When loyal customers are well rewarded, many unanticipated benefits occur. For example, in 1980, just before airline frequent flier programs became ubiquitous, passengers would tolerate an average delay of approximately 30 minutes before switching airlines. In 1990, after most passengers had joined at least one frequent flier program, the tolerance to delay shot up to an average of 3.5 hours, an increase of 7 times.

Whole new markets exist for card issuers who use smart credit or debit cards combined with a loyalty application

Smart cards can integrate whole new categories of loyalty programs which could not be put on a credit card in the past and that relied heavily on central processing, direct marketing databases and mailings.

The complex database marketing process becomes less expensive and much simpler to manage when recency, frequency and monetary value (RFM) behaviour data is placed directly in the smart card, where it can be dynamically updated at each transaction. Direct marketing software in the retailer's point-of-sale terminal interacts with behaviour data in the card, so targeted messages can be provided to customers in real-time right at the point-of-sale terminal, at very low cost. As the rules for giving out rewards are managed on the card and in the terminal, the cardholder can take advantage of them immediately. So additional costs related to processing and handling redeemed rewards are also eliminated. By moving the direct marketing function down to a simple payment terminal, the system becomes far more cost effective and can be applied to small value purchases at smaller retailers.

Loyalty punch cards, punched or stamped by the retailer at each visit, are a perfect example of the type of program that can be migrated to smart payment cards using RFM behaviour data. By storing distinct RFM behaviour parameters for each punch card, a smart payment card can easily carry dozens of programs with little memory overhead. For example, a fast food restaurant can instantly offer a free sandwich at a cardholder's fourth visit in the same month, while a music store might offer a free CD for a pre-stated sum spent on cumulative purchases, with both programs running concurrently on the same card along with many others.

For the cardholder, a smart payment card that can store several electronic punch cards means less wallet clutter and faster accumulation of benefits. It also means the cardholder doesn't have to worry about forgetting to present a paper loyalty card, since the electronic version of the card is automatically 'punched' each time the smart card is used for payment.

For the retailer, it means more efficiency, increased security (cashiers cannot punch the card more than once in the same transaction) and lower costs than with traditional loyalty programs. Also, usage can be monitored at the individual transaction level, which is impossible with paper cards. Most importantly, the ability to run the program on payment cards already issued by a bank means the retailer does not have to issue his own cards. All the headaches and costs of trying to keep a private card base alive and active can be avoided.

For the card issuer, it means a more competitive product, the ability to move from price-based competition to value-based competition and the ability to generate new revenue streams through the added value provided to retailers and cardholders. The positive overall business case also means that loyalty technology can help finance the new infrastructure that is required for smart payment cards.

Changing the rules of the game

With private and co-branded cards, the retailer's loyalty program is completely linked to the card. What tends to happen when a retailer's loyalty program becomes linked to a payment card? A disproportionate share of transactions moves to that payment method. Retailers who issue private store cards tend to have a higher share of transactions paid using the private card. This is a general trend that can be verified in many market sectors. Of course, retailers who issue punch cards see a much higher share of transactions paid by cash and cheques.

Any tool that helps banks effectively move from price-based competition to value-based competition immediately generates significant market interest. Today, payment cards are increasingly linked to loyalty programs which give frequent flier miles for free airline tickets or gift certificates, discounts and other advantages. Most banks offer such cards, so the competitive advantage that the first co-branded card issuers enjoyed is now shared by all industry players. A level playing field means card issuers tend once again to compete primarily on price.

Existing loyalty and incentive programs are generally 'mileage' systems that give *X* points or miles for every pound or dollar spent. In many cases, customers gain the same number of points regardless of their purchase history. Light buyers earn proportionally the same rewards as heavy buyers. This is primarily due to complex and costly database marketing procedures. Smart payment cards provide new possibilities for differentiation by making the mileage program more interactive through instantly available customer statements, printed at each transaction rather than being sent by post several times a year. Even more exciting is that in addition to mileage programs, new benefits that are too difficult or costly to implement with magnetic strip cards can now be provided to customers. Magnetic strip cards were sufficient to establish the large private payment card market. They were also good enough to address the even larger co-branded card market. Now, in order to go to the next step and combine the far larger punch card market with payment cards, chip technology is required.

Early adopters of game-changing technology often enjoy the opportunity to create new brands and lock in partnerships with key service providers. Imagine that ABC Bank creates a new brand, 'Instant Rewards'. The brand is placed on the bank's smart payment cards so they can be easily distinguished from existing cards. Retailers place the 'Instant Rewards' brand on their door, letting cardholders know that they will receive a free item after a certain number of visits or amount spent at that store. Imagine now that ABC Bank lets other card issuers (phone companies, for example, or transport authorities) carry the 'Instant Rewards' brand on their cards, for a fee. ABC Bank will probably try to sign long-term exclusive agreements with major franchise chains, locking up key portions of the market. During prior shifts in technology, in the payment industry as well as in many others, the first organizations that succeeded in building these types of coalitions enjoyed a competitive advantage for a significant period of time.

The same will happen again with the move to smart cards.

Throughout this century, new game-changing technologies have repeatedly managed to supplant existing infrastructures, replacing them with new, improved business models. Smart cards have the potential to redefine payment industry rules and bring about exciting transformation in the types of services offered to customers. This can happen if smart card technology proves itself capable of delivering functionality that offers much more than incremental benefits over existing methods. Smart card loyalty has the potential to be such an innovation.

Transferring dozens of punch cards into a single, smart payment card provides far more benefit over existing methods, especially to the franchisees and smaller retailers that constitute the vast majority of retail outlets. This is game-changing technology in its most classic form.

Gaining market share at the moment of purchase: three fundamental concepts

1. A new card product can substantially increase its chances of success by simply addressing the core motivations of each of the main participants: banks want to increase their share of transactions; retailers want to increase their share of profitable customers, and customers want access to high quality products and services at the lowest price. The most successful payment cards satisfy all three requirements.

2. Early smart card pilots were all single-function implementations, primarily electronic purse. Today, the vast majority of financial institutions are now considering multifunction implementations combining a traditional payment mechanism like credit or debit along with a customer loyalty function.

3. Loyalty and frequent buyer programs are mature applications that have already been linked to hundreds of millions of magnetic strip payment cards since the 1940s. Private cards were the first to be addressed by banks and financial institutions. This market is big, but it is also mature. The market for co-branded cards then became even bigger, but today it too is mature. Punch cards, the biggest loyalty program market, are next to be integrated with payment cards.

2

TRANSFORMING THE SUBSTANCE OF MONEY

❖

Form has an affinity for expense, while substance has an affinity for income.

Dee Hock, Visa founder[5]

The substance of money has not changed very often throughout human history. Money has evolved from the literal (gold and silver coins) to the representational (printed paper money and cheques) to the virtual (electronically processed transactions).

Changes in the substance of money coincide with substantial changes in society, culture and government. Moving to a new type of value exchange, like printed paper money, requires deep, widespread consensus. The expected benefits of a new payment method must be clear, easy to understand, and represent a substantial improvement over prior methods. If the new method is simply a change in *form*, rather than a change in *substance*, it will most likely not take hold. Lacking widespread consensus, it will either disappear or, at best, be relegated to some specialized use at the fringes of mainstream commerce. Changes in form tend to be expensive and risky for their champions, whereas changes in substance tend to generate new income streams.

Electronic – or virtual – money already exists. Money became virtual once banks began communicating transaction information through computers over 25 years ago. In a relatively short period, electronic money became a permanent part of mainstream commerce. Taking the long view, the arrival of smart cards is simply a continuation of the trend that began when banks first placed magnetic strips on their credit cards.

Money has already undergone three stages of fundamental transformation, from literal, to representational, to virtual. Today, smart cards are poised to trigger the fourth stage of substantial change. Money is about to become intelligent.

Substance has an affinity for income

American Express was founded in 1850 as a freight express business. In 1882, in response to the postal money order which was causing demand for the company's cash shipping services to decline, American Express began offering its own 'Express Money Order' service. It was an immediate success. The company jumped on the opportunity and began selling the product at railroad stations and general stores nationwide. This nudged American Express in the direction of becoming a financial services company.

Several years later, American Express president J.C. Fargo was in Europe on holiday, where he discovered the difficulty of converting letters of credit into cash. On his return, the company created an elegant solution which simply required a signature upon purchase and a countersignature upon redemption. This quickly became known around the world as the 'American Express Traveller's Cheque'.

The traveller's cheque presented an unexpected bonus for American Express. In between the time when a cheque was purchased and the moment when it was redeemed, many weeks and even months would go by. This created a comfortable cash cushion. The company sold more cheques each month than it redeemed. American Express had inadvertently stumbled upon the hugely profitable concept of 'float'.

In what started as an incremental step toward solving a traveller's difficulties, American Express had become a fully-fledged financial services company.[6]

These were the early days of the second stage of fundamental transformation of money, from literal to representational. Companies that participate actively in such an important transition period benefit far more than what would appear to be their fair share. And their resulting leadership position turns out to be long-lasting. The opportunity to secure a long-term competitive advantage is again available today for those few, initial financial institutions that help money move into its next stage of transformation.

Using smart credit and debit cards as electronic cash

The most mature smart payment card market is in France, where the banking industry had all cards converted to chip by 1992. Today, the French debit card system is the world's most efficient. Fraud is virtually non-existent. Transaction costs are ridiculously low. The cards can be used anywhere, even at parking meters and to pay motorway tolls for very small amounts, as low as 10 francs. Because the overall system is so efficient, foreign Visa and MasterCard credit and debit cards enjoy the same level of acceptance.

A US smart card consultant visiting France recently commented that the car park we were pulling out of would be an ideal situation for an e-purse, since it was silly to still be handling change for such things. I pulled out my Bank One Visa debit card, inserted it into the machine and paid 12 francs. The consultant was amazed. Word got back to the US and I had several calls from top-level people at one of the world's largest consulting firms. 'Wait, what are you saying? You're doing micropayments in France with your American magnetic strip bank card?'

You can use your card at any McDonald's restaurant in France. Even small one-aisle grocers have payment terminals. Most transactions are off-line, so all the customer does is type in their PIN code. There are no receipts to sign. The whole process typically takes a few seconds, including the time it takes to punch the terminal's keys.

The banking community can go very far in winning market share away from cash and cheques simply by making its current payment products more efficient. A chip debit or credit card will open huge new markets without the hassles and risks of creating a completely new paradigm like an e-purse card that has to be reloaded at a teller machine whenever it's empty. Examples in France of making micropayments with a common debit card are striking.

France's smart payment card implementation is by far the world's most successful. Contrary to e-purse projects being launched all over the world, the French system does not represent a paradigm shift. It is an evolutionary improvement that uses new technology to gain leverage in the financial industry's core payment processing competencies.

Table 2.1 compares two leading smart payment card programs, France's Carte Bancaire debit system and Belgium's Proton e-purse. Proton is generally considered today to be the most successful and mature e-purse implementation. Statistics show acceptance levels two years after nationwide launch.

Any way the statistics are analysed, the French debit card system can be shown to enjoy a very healthy level of acceptance among retailers and cardholders. There are proportionally 4.5 times more places to use the card

Table 2.1 Comparison of smart debit card and e-purse

	Smart debit card	*e-purse*
Card program	Carte Bancaire	Proton
National test market	France	Belgium
Population	50 million	10 million
Cards issued 2 years after launch	21 million (42% of pop.)	4 million (40% of pop.)
Retailer terminals installed	520,000	23,000
Terminal penetration ratio	1 for every 96 inhabitants	1 for every 435 inhabitants
Total purchase transactions during the first 2 years after launch	2.5 billion (119/cardholder)	30 million (7.5/cardholder)
Transactions per month	130 million (6.2/cardholder)	2.3 million (0.6/cardholder)

(1 terminal for every 96 inhabitants, versus 1 terminal for every 435 inhabitants). Over the first two years after launch, French debit cards generated 16 times as many transactions per cardholder as Belgian e-purse cards during Proton's comparable post-launch period.[7] E-purse transactions started slowly but are increasing. After two years of operation, the 16 to 1 ratio had dropped to 10 to 1. It is questionable whether e-purse alone will be able to achieve stronger card usage. Adding a smart debit or credit function to the card along with a loyalty application would go a long way toward achieving a significant increase in acceptance.

Public acceptance of smart cards

It's amazing how fast smart cards became totally accepted in French daily life over a very short period. I moved to France in 1989, just as chip cards were gearing up for national deployment. There was still lots of talk at the time about whether or not the cards would really take off. 'There are too many problems', people would say. 'My card was full the other day and I couldn't do any more transactions,' someone would add. Or, 'There are too many terminals that kill the chip. They'll probably never get the terminals to work right.'

These and many other problems were eventually solved and the trials were made to function properly. By the end of 1992, all French bank cards had been converted to chip.

I still keep a bank account in Boulder, Colorado, my home town, and use my Bank One Visa debit card all the time in France. Sometime around 1996 I began noticing that retailers would occasionally become confused when I handed them my card. They might notice something different with the card, maybe that it lacked a chip, but would go ahead and try to insert it in the chip reader anyway. Once it dawned on them that it was a foreign card, they would revert to the old card swipe wrist movement and complete the transaction. That was a mere four years after all French bank cards had chips. At first I found this a little curious but it didn't strike me more than that. It didn't happen very often. Perhaps these retailers were young, or new to handling cards.

Today, in 1999, not a week goes by that this does not happen to me. It's become worse. Cashiers tell me they don't accept that card because it doesn't have a chip. 'Oh yes you do. I bought lots of things here last week and you were happy to use my card.' A manager is sometimes called over. 'Yes, it's OK, it's a foreign card', the manager will explain to the cashier. 'You have to use the magnetic strip.' Wait, there's more. Retailers in France have lost the card swipe wrist movement. They often don't swipe fast enough, or they swipe too fast, or they swipe the side of the card that doesn't have a magnetic strip. They rub the card on their sleeve (which always makes me cringe) then try swiping it again.

Now I've become used to this. When someone looks confused I explain calmly that the card has to be swiped. When they cannot get the wrist movement right, most retailers are happy to let me take the terminal and swipe the card through myself.

In less than seven years, people have so well adapted to using chip cards that nobody thinks about them anymore. The best designed technology disappears and lets people get on with their jobs. Persuading French retailers to go back to swiping cards would prove to be very disruptive.

I hear people in the UK say that the British will never adjust to entering PIN codes at the terminal. People in other countries have said that retailers and older people are going to have a difficult time adjusting to the card being inserted chip side first rather than being swiped. When I first began developing loyalty marketing concepts around smart cards in France, people would say 'those are American marketing ideas that would never work in France, why don't you go back to the US and try selling them there'. Several years later, presenting the mature software products in the US, some people would say to me, 'I'm not sure American retailers and bankers will catch on to all this, it might work fine in France but over here it's different'.

People who have spent their careers bridging cultures are familiar with these complaints, no matter what industry they work in. Saying that something will not work here, simply because here is here and there is there, lacks sound logic. Products certainly have to be adjusted to each marketplace, which requires a close, open, constructive exchange of information between people with perhaps very different backgrounds. A blanket judgement based on poor logic makes it far more difficult to enter into a constructive discussion of what adjustments are required in order to make a technological product better suited to a particular culture.

E-purse is a change in the form of money, not in its substance

General purpose electronic purse systems like Visa Cash, Mondex and Proton are built to closely resemble cash. Money is withdrawn from an automatic teller machine-like device and stored in the card's chip. Each time the card is used for payment, money is taken out of the chip and electronically stored in the retailer's payment terminal. Once the money is used up, the card must be reloaded again. E-purse is truly presented as a straight substitute for cash, a new form of money. Mondex goes the farthest in this direction by making transactions completely anonymous, like cash, keeping no trace of where and how a card is used for payment. General purpose e-purse cards have not achieved sufficient market acceptance precisely because they are merely a cash substitute, with unclear additional benefits over pocket change. Since it is a change in the form of money and not in its substance, we can expect to see the e-purse have trouble in the mainstream but possibly find some significant success in vertical markets at the fringes of mainstream commerce. This is precisely what we see with the use of the electronic purse in closed environments like launderettes, cinemas, campuses and payphones.

Statistics comparing smart debit cards to e-purse schemes show clearly how much easier it is to persuade retailers and customers to adopt a new payment card product if it does not require too great a change in behaviour and if it provides very clear benefits that all parties recognize easily. For most retailers, the e-purse benefit of reducing the costs of handling change is a concept that appears quite theoretical. That's even more true for customers. How big a problem is change in your pocket? Enough to endure yet another paradigm shift?

The electronic cash pilot in Manhattan ended in 1998 after one year of tests jointly run by Citibank, Chase Manhattan Bank, Visa and MasterCard. Sponsors of the trial say they learned valuable lessons about technology and usage patterns, even though the trial was not a huge hit with customers or retailers. Citibank and Chase issued a total of 100 000 cards loaded with electronic cash. More than 600 retailers were recruited to accept the cards for payment. After 14 months, most consumers never reloaded their cards with electronic cash and two-thirds of the retailers ended up abandoning the trial.

Vice-President Carol Lockie of Visa International said the trial was hampered because residents of the Upper West Side, where the trial took place, could not use their smart cards when they went to work in other parts of Manhattan.

The trial was hamstrung by the lack of an attractive value proposition. After the trial, many bankers and smart card experts came to the conclusion that consumers need a financial incentive to use the cards, such as those provided by loyalty or reward programs.

Dudley Nigg, Executive Vice-President of Wells Fargo Bank, speaking at CardTech/SecurTech 1998, said, 'In Visa Cash and Mondex, we still don't have a really viable loyalty proposition – and that's one reason why smart card pilots haven't been successful'.

A general point of view following customer and retailer reactions to numerous e-purse trials in the United States and elsewhere is that people don't need an e-purse. It's fairly easy to write a cheque, hand over cash, or pay with a credit card.

Harvey Rosenblum, senior vice-president and research director at the Federal Reserve Bank of Dallas, says that 'the existing payment mechanisms that people know and trust work very efficiently and are fairly cheap'. Rosenblum thinks smart cards will appeal to individual customers when they offer perks and rewards to frequent customers.[8]

One big problem is that smart cards and e-purse have become one and the same thing in the industry's perception. Smart cards equals e-purse and e-purse equals smart cards. This is unfortunate. Smart cards were originally used with credit and debit payment instruments, e-purse is relatively new. Without actually putting cash on a card, as an e-purse does, smart credit cards can be very effective in taking market share away from cheques and cash. The French banking industry has shown how this can be done.

Advocates like Visa are now pushing not the single-purpose cards like the ones in the New York trial, but multifunction cards that combine a payment method like credit, debit (or perhaps electronic cash), and loyalty programs to reward customers who return to the same merchants.

E-purse in a closed environment

E-purse as a bank card product has had a hard time justifying its existence. But e-purse used in a closed environment has proved to be far more useful. Prepaid telephone cards in many countries have achieved very respectable adoption rates. French cinemas have been issuing reloadable e-purse cards since 1987, to the tune of 100,000 cards per year. Club Med and countless other companies, universities and college campuses, use them as a way to make paying in a closed environment easier for customers.

How does a closed environment e-purse fit in with a bank's strategy? The core issue is to determine who manages the float. Telephone companies have become accustomed to holding on to the customer's money for quite long periods before the card is actually used up. In France, experts have estimated that a full 18 per cent of prepaid amounts are never used! The amount sits as residue on the card for an indefinite period of time. Cinemas have also grown comfortable with the practice of managing the float on their cards. Financial institutions will have great difficulty acquiring and managing the float generated by a closed environment e-purse.

During the mid-1990s, several large e-purse projects linking telephone companies with financial institutions were under preparation, in the United States and several other countries. Most of these discussions never resulted in a tangible implementation. The main problem concerned float.

'I'm a financial institution,' says the bank, 'it's my role to manage the float, plus my business case depends on it. Why else would I issue e-purse cards?'

'Perhaps,' says the telephone company, 'but those are my payphones out there, I can't justify the cost of upgrading them if I can't manage the float.'

E-purse cannot on its own deliver the business case justifying a bank's move to smart cards. The business case must come from leveraging the bank's core expertise with credit and debit cards, improving those core services by adding new functionality and reducing transaction costs through efficient technical procedures. Even though a general purpose e-purse cannot be seen as the heart of the bank's business case for smart cards, a closed environment e-purse might eventually prove to be a useful value-added service that makes the card more attractive to customers. In the same way that using the card to secure access to buildings or to private networks might one day prove to be additional functions that people would like to have on their credit cards.

In the meantime, banks can leverage their current credit and debit cards to win market share from cash and cheques, simply by using low cost smart cards with a credit function and electronic punch card capability. This is powerful, inexpensive, simple to use and easy to operate.

Visa's Relationship Card

Visa is creating a new generation payment product based on smart cards. They call it a 'Relationship Card'.[9]

You carry a Banking/ATM card to access funds on deposit, a credit card for purchasing and cash for small value expenditures. A Relationship Card will do all that and more, enabling you to maintain a portable relationship with your bank, with access to all your accounts, at any time. In addition to multiple account access, the card will let you obtain account information, transfer funds between accounts and load currency onto the stored value component, eliminating the need to carry coins or exact change.

Loyalty programs – like airline frequent flier, rental car VIP and video store frequent renter programs – reward customers for incremental card and/or product usage. Thanks to its microchip intelligence and macro capabilities, a Relationship Card will track and store data for loyalty programs automatically using transaction information as each purchase is made. Imagine having real-time updates on loyalty point balances and instant access to rewards as they become available ... 24 hours a day, all over the world.

Just as you expect communications flexibility from telephone service providers, you expect banking flexibility from your financial institutions. You want access to customized financial products and services through electronic devices such as screen phones, personal computers, personal digital assistants, shared ATMs, point-of-sale terminals and other future devices. You want help in managing your various accounts, guaranteed security and control over your personal finances. Relationship Cards will deliver all this and more.

The power and intelligence of the microchip will enable a Relationship Card to be more than just a financial tool. It will also be able to store a myriad of personal information that you might otherwise carry in your wallet. It could act as an all-in-one repository for medical and dental insurance information. You decide what information you store, and a security system designed into the chip means data stored in the cardholder information part of the chip can only be accessed by you or those whom you designate.

Money is poised to become intelligent

In the past, money has evolved to become more and more symbolic. Paper money and cheques are symbols void of intrinsic value (the paper costs nothing). Virtual money takes the same concept further by using electronic bits as the symbol for value. All forms of symbolic money are backed up by governments, legal bodies and financial institutions.

Smart cards introduce a brand new element. The bits are of course a symbol for money, that's not new. But now money in the form of a smart card payment instrument is able to store and understand information concerning itself, how it has been used in the past, where it has been used, when, and how often. Suddenly, money becomes intelligent by its ability to watch and monitor itself and its user, reacting to stimuli that it and its user have generated. It can cause things to function differently if the card has been used frequently over a period of time, or if it is the first time it is used at a particular place. It can trigger warnings, and reminders, either to its user or even to itself.

Many readers will recognize this as a classic feedback loop, the process of using the output of a device as new input for the same device. Feedback loops are at the heart of many of mankind's most useful and revolutionary technologies. James Watt created a regulator for steam engines that used a feedback loop to stabilize the motor at a constant speed of the operator's choice. It launched the industrial revolution. Telephone engineer H.S. Black, working at Bell Laboratories, created the first electrical feedback loop in 1929 in his search to improve amplifier relays for long-distance phone lines. This paved the way for the invention of vacuum tubes and transistors, the building blocks of the information revolution.

Feedback loops appear naturally in biological processes. Neurons, DNA cells, higher organisms, all depend heavily on feedback loops. In fact, not a single example has been discovered of a biological process devoid of feedback. It also appears in complex social structures and in economic theory.

Once they were discovered, feedback loops have been introduced in more and more human engineered objects, resulting almost always in spectacular innovations. Taking the long view, using smart cards to create a feedback loop for money was inevitable.

Feedback loops allow an entity to become self-governing, self-adjusting and capable of adapting to many different and unpredicted stimuli. When layers of feedback loops are piled on each other, resulting in the complex organisms encountered in life, they allow entities to learn and adapt to the world in a way that we intuitively consider to be alive. Consciousness itself, emerges from layers upon layers of feedback among several billion neurons that recognize nothing more than two states, charged and discharged. When feedback is applied to very large numbers of ridiculously simple entities, the overall system becomes excruciatingly complex. When systems become sufficiently complex and interconnected, they self-assemble into a new, higher order, that is more than the sum of its parts. Human engineered entities are just beginning to become complex in a similar fashion, as they integrate layers upon layers of feedback. The global telephone network is one example of a complex system that in some biological sense could almost be compared to a living entity.

As the substance of money is transformed and becomes increasingly intelligent, it will adapt to provide its user with greater services and convenience. Throughout humankind's relationship with technology, whenever human engineered objects were made aware of themselves, floods of innovations appeared that were completely unanticipated. This is about to happen with money.

Transforming the substance of money: three fundamental concepts

1. Money has already undergone three stages of fundamental transformation, from literal, to representational, to virtual. Today, smart cards are poised to trigger the fourth stage of substantial change. Money is about to become intelligent. History shows that companies that participate actively in such a major transition period tend to secure a long-term competitive advantage in their industry.

2. Money in the form of a smart card payment instrument is able to store and understand information concerning itself, how it has been used in the past, where it has been used, when, and how often. Suddenly, money becomes intelligent by its ability to watch and monitor itself and its user. This is a classic feedback loop of the type that has produced many of mankind's most useful and revolutionary technologies.

3. The banking community can go very far in winning market share away from cash and cheques simply by making current payment products more efficient. A chip debit or credit card will open huge new markets without the hassles and risks of creating a completely new paradigm like an e-purse card that has to be reloaded at a teller machine whenever it's empty.

3

REAL-TIME CUSTOMER RECOGNITION

❖

If you are not too long, I will wait here for you all my life.

Oscar Wilde

Here are some of the things that my chip equipped credit card will soon be able to do for me at a retailer's point-of-sale terminal:

- I receive a free item at my fourth visit to the same store in the same month.
- Coupons (that I got from the Web) are automatically taken from my card for payment when I purchase the couponed item.
- For every four items I purchase of a specific brand, I receive a coupon for a fifth item free.
- I receive a $20 gift certificate if I have significantly increased my purchases at a particular store.
- At my first visit to a store, I'm recognized as a potentially very valuable customer (because I spend a lot in this retailer's category, but not currently at his store) ... and instantly receive a gift certificate.
- The store manager comes and talks to me when I don't shop at his store as often as I used to.
- I can choose to go on the retailer's mailing list for monthly specials and other information that is adapted to my behaviour. The

retailer's monthly letter is sent to me by e-mail.

As a retailer, why can't I instantly know that a customer's behaviour has changed dramatically? It would be great to know that the customer standing in front of me has just doubled his monthly spending and that his projected lifetime value is now over $500,000. I might want to offer that customer a special gift, a bottle of wine or a ticket to the opera.

On a more defensive note, if a customer has reduced her spending recently, I might guess that she is probably spending far more at my competitor's store. Since she's standing right in front of me now, maybe I can do something about it.

What if a new customer begins shopping at my store, and I have a way of instantly knowing that she currently spends a lot in my retail category? Wouldn't I want to look for ways to treat this customer with great care, even if she doesn't currently spend much in my store?

Here is the real-time message, according to Regis McKenna:[10]

> New consumers are never satisfied consumers. Managers hoping to serve them must work to eliminate time and space constraints on service. They must push the technological bandwidth with interactive dialogue systems – equipped with advanced software interfaces – in the interest of forging more intimate ties with these consumers. Managers must exploit every available means to obtain their end: building self-satisfaction capabilities into services and products and providing customers with access anytime, anywhere.

Avoid mileage programs – or do them right

When one mentions loyalty, many people instantly think of points or miles like those earned through frequent flier programs. Mileage type programs generally allow customers to acquire points across many different retailers. Points can then be used for payment at the same or other participating retailers. When mileage points are given to a cardholder, the retailer immediately owes the program operator the value of those points. If, for example, points are given at the rate of 2 per cent of each transaction's purchase value, this is really just another way of giving a discount of 2 per cent. Rather than providing it directly to the customer, the discount goes to the program operator who manages it for the customer, in exchange for a commission.

Virtually all private and co-branded card programs function on mileage. Programs are so similar that the Web literally crawls with tables comparing dozens of credit card features like where you can use them for mileage, what the value of the miles is, how many miles are required for a minimum reward, and so on.

An easy to remember tip about mileage programs is avoid them. If you can't, be extra careful to do them right because it is so easy to do them wrong.

If the program operator is capable of adding significant value to the accumulated points, the system makes sense for everyone. The airline industry created the mileage concept in 1981, when American Airlines launched the AAdvantage frequent flier program. Miles are good for free tickets and upgrades, which have a very high perceived value, but represent a low incremental cost to the airline. Airline seats are 'perishable' – once the plane takes off, an empty seat can never be sold again. Filling an empty seat with a passenger that pays with miles costs very little in incremental costs ($18 to $54, mostly for food and fuel, according to a 1998 *Consumer Reports Study*[11]) and can even generate additional revenue when the passenger travels with a spouse paying full fare. The huge discrepancy between the perceived value of miles and their actual cost means that customers can acquire attractive benefits in a reasonable time frame. The economics described here are applicable to all industries that sell perishable items like airline seats, hotel rooms, car rentals, etc.

Successful frequent flier programs have been copied by companies in many other industries, but with far less success. Non-perishable items have a real cost to the seller that is much closer to the customer's perceived cost than to an incremental cost approaching zero. Compared to airline seats, the retailer's cost of a box of cereal is very close to the customer's price, which is why it takes such a long time to acquire attractive benefits with most supermarket and petrol station loyalty programs. Mileage programs in these industries are forced to find a way to create a wider discrepancy between the actual cost of the reward and its perceived value. This is very difficult. Program operators are generally faced with two options: either add meaningful value to the points, or dramatically reduce the cost of the rewards without sacrificing quality. Few mileage program operators have succeeded in either approach.

In most cases the mileage program operator simply acts as a financial clearing mechanism, acquiring points from retailers that issue them to cardholders, and reimbursing retailers for points used for payment in their stores. This is why many retailers view mileage as a complicated discount mechanism that does not necessarily build loyalty to their store. For customers, mileage is seen as a deferred discount that must be accumulated over a long period in order to be eligible for anything worth while.

Not only do mileage programs typically fail to add value to the points they manage, but worse, they succeed in destroying value through huge operating costs. When discounts offered by retailers are converted to mileage points managed in a central computer, costs like monthly statement mailings and voucher processing eat up a large portion of the discounts that were originally supposed to be for the customers' benefit. This is a problem that plagues mileage programs in all industries. It becomes particularly unmanageable in industries that don't enjoy a high discrepancy between the perceived value and the actual cost of mileage points.

If you are currently a mileage program operator, here's a simple way to know if you add sufficient value to the points or not: *ask retailers if they would prefer that the points they issue be valid only in their store.* In other words, would they prefer that discounts they offer be used by customers to purchase additional products and services in their stores? If the answer is yes, you have a clear indication that retailers feel that your program does not add value to the discounts they can directly provide their customers. In fact, a common complaint heard among retailers participating in poorly developed mileage programs is that those programs are a loyalty mechanism for the card operator, not for the retailer.

Retailers have been successfully building customer loyalty long before multi-retailer mileage programs were invented. Punch cards, for example, constitute a mainstream solution for many retailers, and they lack the drawback of being expensive and cumbersome to deploy over large numbers of participating stores. A mileage program operator's added value must be very significant in order to switch retailers away from proven loyalty methods that don't require complicated and expensive third-party points management.

Most card issuers will place greater priority on the electronic punch card approach as opposed to the multi-retailer loyalty points mileage approach. Electronic punch cards provide significant value to the retailer and to the cardholder, who enter into a 'friction free' relationship in which the customer's loyalty is given in exchange for hefty discounts from the retailer. No middleman manages the retailer's discount, deferring it for future payment once the customer has acquired a sufficient number of points (and taking a large commission in the process). Punch cards dramatically reduce the program operator's role, since redemption and clearing of points across numerous retailers is no longer required. If your goal is to make your card program simple, scaleable, easy to install, deploy and operate, you should definitely consider this approach.

Because a mileage program operator must add substantial value to collected points, we believe mileage type loyalty is best suited to programs run for *clubs, affiliations and other-cause related organizations.* Such organizations can use the mileage points for specific causes that customers hold dear, the emotional element adding significant value to the mileage points. Cause-related mileage programs are an ideal add-on to an electronic punch card loyalty system. Punch cards address the needs of all retailers and cardholders, while mileage points can be used to address the needs of specific segments of customers that would like to participate in cause marketing programs, in addition to benefiting from punch card discounts.

Children's Heros is a cause marketing organization that uses mileage points to raise funds for schools in the US. Chris Hutcherson, President of Children's Heros, first launched a paper voucher version of the program in 1992. Families purchased the vouchers at face value from their schools and used them to pay at participating retailers. School mothers manually collected and processed the vouchers, reimbursing retailers for their face value minus a rebate. A percentage of the rebate went to the school, the rest went to pay for such things as the individual child's tuition, field trips, classroom aids and scholarships.

Over a period of six years, the program went from one State to 44, from 385 schools to 11 000, from 40 000 families to 4 million, from 8 major retailers to 265. From $24 million in sales, generating $1 400 000 for schools, the program reached $1,600 million in sales and generated a whopping $107 million for schools.

Chris Hutcherson says that parents see participating retailers as:

> corporate citizens and community leaders that care about and do good for our kids, our families and our schools. The love parents have for their children is the single most powerful force on earth. Our children stand at the center of our universe. To nurture one's child is each parent's duty, but the education of every child transcends boundaries required of parents only. In the truest sense of duty to the communities in which we live, the well-being, the education and the future of the children on this planet is the responsibility of us all.

Talk about emotional added value. Many parents would not hesitate when faced with the choice of pulling out a credit card that earns frequent flier mileage and one that earns money for their child's school. If your mileage program doesn't generate this level of emotional value, you'll need to fix it at the core.

Guarantee that your co-branded credit card program builds loyalty to the retailer

Companies as different as General Motors and Toys R Us have launched scores of co-branded credit cards over the last ten years. These are cards with the General Motors or Toys R Us brand on the front, next to a Visa, MasterCard or AmEx payment brand. Some programs have been quite successful while others have had very poor results. Bad press from some retailers has caused many now to lump all co-branded card programs together and dismiss them as a failure.

A common complaint concerning programs that have not succeeded is that the retailer perceives the co-branded card as building loyalty to the issuing financial institution as opposed to loyalty to the store. This has prompted some retailers to go so far as to ask the card issuer to pay licence fees for the use of the retailer's brand.

Most co-branded card programs work the same way. The cardholder receives 1 point for every dollar spent with the card, anywhere the customer shops. Once a significant number of points are collected the customer can use them to buy things at the retail chain whose brand is on the card. Sometimes the retailer adds bonus points for shopping at that retailer's stores, for example by doubling the points obtained at each transaction.

Who is the primary loyalty beneficiary in this partnership? The financial institution or the retailer?

If a large majority of the points are obtained shopping outside of the retailer's chain, then from the retailer's perspective points simply represent another payment currency. Retailers are referring to this problem when they say that their co-branded card primarily takes transactions away from other payment methods and therefore builds loyalty to the financial institution and not to the retailer.

On the other hand, if customers primarily obtain their points by shopping at the retailer's stores, then the card would be a very valid reward mechanism totally dedicated to customers loyal to that retailer.

Another potential problem from the retailer's perspective concerns the impact cardholders have on revenue lift. If too few customers are cardholders, it's clear that the impact on overall revenue will probably be small. What percentage of total revenue do cardholders represent? Anything under 10 per cent will have too little impact on sales to make a noticeable difference.

In some cases this might not be a problem at all and this type of co-branded program might be a complete success. The General Motors card is a good example. Every time the GM card is used 5 per cent of the transaction amount is credited to a special rebate account that can be applied toward the purchase of a new car. When ready to buy, the cardholder calls a freephone number and requests a certificate that can be redeemed at any GM dealership. Five years after launching their MasterCard credit card, General Motors had sold over 1 million cars to customers using rebates from the program.

The retailer and the financial institution both need to clearly understand their common objectives concerning loyalty. They also need to understand the potential impact of their choice of program. The specific mechanics of the program must be defined, taking overall loyalty objectives into account. Otherwise, someone will more than likely feel they are producing the lion's share of value in exchange for a small portion of the benefits.

Distributed control

When NASA began landing unmanned roving devices on the moon, they quickly discovered that centralized control from Earth did not work very well. The one-minute delay between an Earth-based central command station and a robot about to go over a cliff meant that robots had to be autonomous. Intelligence had to be placed directly in the device to allow it to make decisions quickly without detailed guidance from Earth.

It turned out that troops of small, autonomous robots built for specific chores, like searching for minerals, or preparing landing sites, were cheaper and more robust than a single, general purpose robot centrally controlled from Earth. These robots can be built quickly and easily from off-the-shelf parts. They are inexpensive to launch. Once released, they can perform many useful services without the need for constant supervision.

Scores of intelligent devices will become part of our lives, smart doors that turn off the lights when you step out, smart stereos that dynamically adjust the balance based on where you are in the room, smart vacuum cleaners that come out when no one is home, randomly clean up for a while then discreetly go back to their place under a bed or in a corner. These types of household devices all have one thing in common: they are not linked to a central command system, a sort of house brain that manages every little detail. Each device has its own (limited) intelligence. It does one thing tirelessly. If the smart vacuum cleaner breaks down the dust might start to collect, but the other devices keep on functioning.

Individual devices might send messages periodically to a unit that monitors and reports on potential problems. The vacuum cleaner might send a standard e-mail message once a day, telling how many miles it has swept up since it went into service, how much dust is collected in the bag, as well as information on potential problems like a wheel being clogged up. In this case, the central unit acts as a monitor, not as a controller. It adds a layer of value on top of individual autonomous devices without getting in their way and without creating bottlenecks.

Kevin Kelly, executive editor of *Wired* magazine, says:[12]

> the surest way to smartness is through massive dumbness. The surest way to advance massive connectionism is to exploit decentralized forces – to link the distributed bottom. How do you build a better bridge? Let the parts talk to one another. How do you improve lettuce farming? Let the soil speak to the farmer's tractors. How do you make aircraft safe? Let the airplanes communicate among themselves and pick their own flight paths. This decentralized approach, known as 'free flight', is a system the FAA is now trying to institute to increase safety and reduce air-traffic bottlenecks at airports.

Moving intelligence out to the peripheral is a paradigm that echoes what happened when electrical power first became widely available. At that time, the Sears Roebuck catalogue advertised a general purpose electric motor for household use, complete with dozens of attachments and optional add-ons. The motor could be reconfigured to act as an egg-beater, an ice-cream maker, a clothes washer, power shears, a drill, a saw and even a personal vibrator device, presumably to ease sore muscles. At the time, it was difficult to imagine colonies of little motors built into each individual device. That would have been considered over wasteful and costly.

To many engineers designing complex technology, the logic of giving up central control is upside-down to the generally accepted approach of applying thorough, top-down control to every step of the process. Although distributed control based on a bottom-up design might be counterintuitive, it nevertheless consistently proves to be less expensive, more robust and more effective in many areas of technology as well as in economic and social systems. Electric motor devices is one example, robotics is another. And the Internet is yet another. Huge global computer networks are based on chunks of peer-to-peer networks. Imagine instead a top-down, traditional network on a global scale. Who owns it? The Federal government? A union of nations that adhere to the same trade treaty? You have a problem with your e-mail? Call the Department of E-mail Administration in Washington, D.C.

Centralized control is a residue of the industrial age. Distributed control and real-time response to external stimuli have already become mandatory features in many industries. Smart cards are now bringing this same trend to the payment card industry.

Real-time customer recognition requires distributed control

Centralized loyalty programs creep along, exploring unmapped territories of customer behaviour, under the constant guidance of a central control unit. They routinely encounter numerous obstacles, and can faithfully report that a cliff is appearing on the horizon, is approaching, is at an uncomfortable distance, is here, now. Information is digested long after it is generated. The central control unit can of course change directions and adjust the program in many different ways, but commands take a long time to reach the faraway program. Several days or weeks after the program falls over the edge of a cliff, central control is proud of its ability to provide a detailed, second by second account of the disaster. Moreover, central control will continue to check for vital signs, as the program lies in pieces at the bottom of the cliff, reporting that yes, the program is still lying there and that it hasn't moved much over the past weeks and months. The few times something did move down there might prove that the program is still somewhat alive. On the other hand, it might have been a few good gusts of wind. Nobody really knows.

Centralized customer relationship programs are unable to react in real-time to a customer's behaviour. They send out new commands via direct mail messages that arrive usually after it is too late. Attempts to speed up communications have resulted in tremendously complex and expensive systems that are notorious for their tendency to collapse under their own weight.

Monthly statement mailings, data processing staff, database experts operating sophisticated segmentation software, customer service centres dedicated to statement inquiries... that's the expensive loyalty card infrastructure that is required for a typical centrally commanded program to function properly. To cut costs, the most dramatic thing to do is to cut out one or two mailings a year. This effectively reduces annual costs by 25 per cent or more, but it only makes the problem worse, since it increases the already unacceptable delay between something happening in the field and when the central marketing structure finally responds to it.

Another reason why loyalty programs cannot be designed from the top down, is that each individual retailer's needs are different. This is of course true for large, multi-participant loyalty programs based on mileage, where customers receive one mile for every dollar spent. It is also true for smaller programs limited to a single chain, since each store in the chain has its own distinct set of problems, competitive pressures and opportunities. The individual store manager is usually the person best suited to responding quickly to local market requirements, especially in a franchise environment where each outlet belonging to the chain is owned and managed by a different person.

Robustness and order emerge from systems that are designed from the bottom up. In his book, *Complexity: The Emerging Science at the Edge of Order and Chaos*, Mitchell Waldrop writes: 'Since it's effectively impossible to cover every conceivable situation, top-down systems are forever running into combinations of events they don't know how to handle. They tend to be touchy and fragile, and they all too often grind to a halt in a dither of indecision.' He suggests that to build robust systems, one should use local control instead of global control. Let the behaviour emerge from the bottom up, instead of being specified from the top down. And while you're at it, focus on ongoing behaviour instead of the final result. Living systems never really settle down. The control of a complex adaptive system tends to be highly dispersed. There is no master neuron in the brain, for example, nor is there any master cell within a developing embryo. If there is to be any coherent behaviour in the system, it has to arise from competition and cooperation among the agents themselves.

Visa's design was based on these concepts. Dee Hock, Visa's founder, came to the conclusion that 'it was beyond the power of reason to design an organization to deal with the complexity of the credit card industry, and beyond the reach of the imagination to perceive all the conditions it would encounter – yet, evolution routinely tossed off much more complex organisms with seeming ease'. It became apparent that the organization Dee Hock was creating (later to become known as Visa) would have to be based on biological concepts and methods very different from the traditional hierarchical command-and-control type organizations that Hock says, 'were not only archaic and increasingly irrelevant, they were becoming a public menace, antithetical to the human spirit'. Hock's organization would have to evolve, in effect, to invent and organize itself. Visa was painstakingly designed based on Dee Hock's belief that 'simple clear purpose and principles give rise to complex, intelligent behavior – complex rules and regulations give rise to simple, stupid behavior'. Visa is highly decentralized, with authority, initiative, decision making all pushed out to the periphery of the organization, to the member banks themselves.

Likewise, robust customer marketing programs must be designed from the bottom up. Intelligence must be moved away from the centre and placed where it is most useful, directly at the point of purchase, so the marketing program can become autonomous and respond to customer behaviour in real-time. Real-time technology provides customers with higher quality service and immediate access to benefits, without all the usual administrative overheads. Programs capable of real-time customer recognition cost less and provide better service than those that are not capable of real-time recognition. Smart cards are the key to achieving this.

Some of the first smart card-based loyalty programs allowed a customer's points or frequent flier miles to be stored in the card's chip. This meant that cardholders could know their card balance every time the card was used, rather than having to wait for a statement to be sent by post. By effectively providing a loyalty account statement each time the card is used, the customer immediately knows that the rewards have indeed been correctly added to the card. So they no longer need to call customer service to ask why last week's flight did not appear on the statement just received by post.

Boots the Chemists, a British pharmaceutical chain, issued over 8 million smart loyalty cards to customers in the first year of operation. All 1260 stores were equipped with smart card terminals capable of calculating loyalty points based on a purchase amount, updating the points in the chip and printing out a loyalty statement with the updated points total. The card is estimated to have cost Boots £52 million over 3 years of operation, an investment that was more than offset by a sales increase of over 4 per cent.

Table 3.1 shows Boots the Chemists' investment over the first three years of operating its smart card loyalty program.[13]

Table 3.1 Boots the Chemists' smart card loyalty program

	1997	*1998*	*1999*
Marketing (£m)	8	8	8
Capital investment (£m)	4	–	–
Database (£m)	4	2	2
Other (£m)	3	3	2
Cost of cards (£m)	6	1	1
Total investment (£m)	25	14	13
Total investment as percentage of sales	0.5	0.28	0.24
Sales increase required to achieve breakeven (%)	4.5	2.5	2.2
Actual sales increase observed (%)	4	4	4

In general, loyalty programs that don't use real-time techniques are estimated to typically cost 0.5 to 1 per cent of sales. In other words double or triple Boots the Chemists' investment. Boots spent a total of £52 million over three years. A similar program using non real-time techniques would have cost easily between £100 million and £150 million and would likely have resulted in a smaller increase in sales.

Smart card programs cost a great deal initially, during the first year of operation, due to the cost of the cards themselves. But overall operating costs go down significantly once the program is launched. Traditional non real-time programs cost a lot the first year of operation, then begin increasing in cost each subsequent year, primarily because of huge postage expenses.

Even if we assume that a real-time loyalty program has the same impact on customer spending as a non real-time program, and that the sales increase in both cases is the same, using smart card technology presents less risk than using magnetic strip or other non real-time technologies – the now factor has the ability to better impact customer spending.

The space between the virtual world and the physical world

It is hard to find good English language bookshops in France. They are few and far between and very rarely carry books I'm interested in. A few mail order catalogues used to help out, but even their choice was limited. Then

Amazon.com came along. Now I spend almost $100 each month on books and CDs. Each time I go to Amazon.com's Website, I look up the customized list of books that they recommend to me, based on the books I have bought in the past. I've discovered all kinds of books that I would never have looked at, much less bought, in a bookshop.

The type of real-time service offered by Amazon.com can of course be extended to all kinds of other retail sectors. Web technology is particularly well adapted to the needs of real-time customer recognition.

But purchases from the Web still remain a drop in the ocean in overall retail sales. US consumers spent approximately $2.6 billion on the Web during the 1998 holiday season. That is barely 1 per cent of total holiday retail sales. On-line spending is expected to grow to $108 billion in 2003. That will be just 6 per cent of projected retail sales. Catalogue shopping currently accounts for about 7.1 per cent of retail sales. How much business is the Web taking away from traditional catalogue sales as opposed to real store sales? Does the growth in on-line sales simply reflect how the catalogue market is moving to the Web? Can the Web be used to extend the current market share of catalogue sales? What market share beyond 7 per cent can on-line sales reasonably expect to reach?

Clearly, Amazon.com has shown that the Web provides the ability to implement very useful real-time marketing techniques. Similar techniques would be of great value to retailers in the physical space as well.

Amazon.com and other Web-based retailers are able to offer real-time services because they know the customer. Every time you go to their Website you are instantly recognized. Each customer's database is on-line all the time. The retailer knows who you are, how good a customer you have been in the past and what specific items you usually prefer.

This is precisely the type of real-time customer recognition that smart cards bring to the physical space. Smart cards provide real-world retailers the ability to efficiently access each customer's behaviour information in real time, and offer instant services that can help retailers forge more intimate ties with their customers.

Real-time RFM

One of the loyalty functions most in demand for smart cards is the off-line delivery of targeted advantages such as a free gift or discount after a specific number of visits or money spent by the customer in the same month. Basically, electronic punch cards. This capability requires the processing of smart card level recency, frequency and monetary value parameters (RFM) for each punch card stored in the chip.

Although the RFM customer segmentation method is well known by database marketing professionals, transferring the process to smart cards proved to be non-trivial. Database professionals traditionally use RFM to rank customers based on the recency of their last visit, the frequency of their visits in a given period and the cumulative monetary value of their purchases. For example, customers might be assigned a recency score of 24 if they have shopped within the last 4 weeks, a score of 12 if they shopped between 4 and 6 weeks ago, 6 if they shopped within the last 9 weeks and a score of 3 if they shopped within the last 12 weeks. Similarly, scores will be attributed based on the total amount the customer has spent (for example, a monetary score equal to 10 per cent of the total amount spent) and the number of visits made (say, 4 points for every visit made in the year). The method results in a sorted list of customers based on their overall score.

Once customers are compared to each other and ranked in this fashion, a marketing strategy can be developed that takes into account the behaviour of best versus worst customers. Offers are formulated differently for each group and are sent by post to each customer.

The known RFM method requires access to a full customer database with detailed transaction data covering each customer's purchase history. It was designed to function on a central computer system that processes data including at the minimum the date and amount of each transaction linked to a specific identifier of each customer. The customer's identity is usually known via an ID placed in the customer's magnetic strip card that is swiped at the checkout register, allowing the customer's ID to be linked to the purchase transaction. The transaction can be uploaded instantly (in an on-line configuration) or nightly (in a batch mode).

To a person skilled in database marketing, smart cards do not inherently provide an advantage over other methods that allow a customer's identity to be linked to a purchase. Magnetic strip cards work fine. So do simple paper cards and key-rings with the customer's ID printed in the form of a barcode. Supermarkets have been using barcoded customer cards and key-rings for over a decade. The card is scanned at the point-of-sale terminal, just like an item's UPC code. The customer's ID becomes part of the purchase transaction and is uploaded to the store's server along with the products that were purchased by the customer. There, a marketing database management system can classify and segment customers based on their purchase behaviour. Average purchases on a per household basis can be calculated. And marketing plans can be formulated to encourage specific categories of customers to change their purchase habits. Letters are sent out to selected customers, offering them a free turkey if they spend a minimum of $500 before Christmas. This approach works quite well when the store manager can master the complexities of database marketing. Smart cards are not required, so database marketing specialists have rarely attempted to use smart cards to store and process RFM behaviour parameters.

Although the common approach described above, using barcoded cards, does not rely on real-time methods, many PC-based point-of-sale scanning terminals have nevertheless offered real-time loyalty features for years now. A feature commonly offered since the early 1980s is the ability to use a customer's barcoded card to trigger the terminal to give specific discounts reserved for that customer's specific category. Customers that sign up for the retailer's loyalty program receive special discounts and a monthly newsletter. Some of the most effective features available with modern point-of-sale scanning terminals do not even require a customer to present a loyalty card. After scanning all the purchases in the customer's shopping basket, the point-of-sale terminal goes back and adjusts prices if the overall transaction is over $20. This is how systems handle the dual pricing mechanism used in many US supermarkets, offering convenience store pricing to those customers who just buy a couple items, and discount store pricing to those who buy more.

Some systems are even capable of storing the customer's loyalty points on the same in-store server that provides price lookup information every time an item is scanned at the point-of-sale terminal. This effectively allows some customer information to be available in real time at the point of purchase.

However, server-based customer information does have several drawbacks. In particular, real-time access to a customer's behaviour across multiple stores belonging to the same chain becomes extremely difficult and costly. The type of marketing analysis that segments customers based on their purchase history would quickly bog down even the largest and most powerful servers available today, if analysis is attempted in real time. Centralized systems work fine when rewards are sent out by post, but they prove to be completely inadequate when rewards need to be delivered in real time. A different approach is required, more robust and more easily capable of performing complex marketing analysis on the spot. In the late 1980s, smart card pioneers were already promising that the new technology would one day allow this.

The traditional RFM method could not simply be transferred to a smart card without major modification to the method in its implementation and even in its spirit. It was necessary to recombine smart card technology with the RFM method in a completely new, innovative fashion in order to arrive at a useful result.

The difficulty of using smart cards as a loyalty marketing device is illustrated by the number of companies that have made such attempts but for various reasons fell short of fully using the card's real-time capabilities.

In 1988, a US company, Advanced Promotion Technologies, developed an interactive electronic marketing system delivering targeted promotions, a frequent shopper program, financial services and other information to shoppers in supermarket checkout lanes. Terminals used full motion video and stereo sound to present the consumer with informational messages while touch-screen displays were used for consumer interaction. The system initially planned to use smart cards for the frequent shopper program. The company was very well positioned to leverage smart card technology and create innovative methods to build customer loyalty. Shareholders included packaged goods manufacturer Procter & Gamble and a well known database marketing firm. And yet, in spite of all this, the company never did succeed in using the smart card as a repository device containing dynamically processed customer behaviour information. Smart card technology was finally abandoned as the company moved to a completely on-line architecture requiring satellite communications. The project never achieved the ease of use and cost efficiency necessary for a mass deployable mainstream solution.

HTEC, a UK firm, is a leading supplier of loyalty terminals to petroleum companies. Using magnetic strip read–write devices at the point of sale, HTEC's system is capable of dynamically storing information directly on the card's magnetic strip. Reading information from a card's magnetic strip is easy, you simply have to swipe the card. Writing information is much harder. You have to swipe once to update the magnetic strip and then swipe a second time to make sure the update succeeded. If an error occurred, the card would need to be swiped another two times. This is why magnetic strip cards are better using motorized devices that can automatically swipe the card multiple times without operator intervention. But motorized card read–write devices are very expensive and prone to physical problems like jamming. The overall magnetic strip card system's economics are therefore based on very expensive, motorized read–write terminals and very low cost magnetic strip cards. The business case makes sense in situations where there are few terminals but many cards. Although the customer's total acquired loyalty points are stored in the magnetic strip, other information related to the customer's behaviour, like cumulative spend in a given period, or frequency of visits, must be stored on an external database server.

Other companies have also developed similar techniques using smart cards, but all fall short of placing the customer's detailed behaviour parameters in the smart card's microprocessor chip, and dynamically processing them in real time at the moment of purchase.

A breakthrough was achieved when smart cards were eventually used to create a distributed database, placing behaviour information directly in the customer's wallet and direct marketing software in the retailer's terminal:

- Behaviour data, including RFM parameters, is calculated, stored and updated in real time on a smart card, as opposed to being calculated, stored and updated centrally in batch mode.
- The ranking process at the heart of the RFM method is dispensed with, since in an off-line mode, real-time access to a full customer database is not possible. Each customer transaction must be analysed independently of all other customers, which is contrary to the traditional RFM process.
- The ranking process is replaced by an algorithmic processing at the point-of-sale terminal that reacts in function of each customer's behaviour profile in order to determine whether or not to print a coupon.
- A coupon containing a message or an offer is automatically printed out directly at the point-of-sale terminal based on behaviour data in the card.

This idea was originally conceived as a method to enhance mileage points programs by adding the ability to vary the number of points issued based on RFM behaviour parameters, as opposed to simply adding points to the card as a fixed percentage of each purchase. Upon marketing the system, it was discovered that the process was applicable to another, much wider use: the replacement of loyalty punch cards that are hole punched or otherwise marked at each visit, providing the customer with a free item or service once the card is full of punches.

A single file in the smart card's memory can hold numerous RFM slots, each dedicated to a retailer's electronic punch card program. The process can be made very efficient; a 500 byte card file can store up to 30 distinct electronic punch cards. This very small footprint can be supported by many low cost smart cards which were originally intended as single-function credit or debit cards, and which effectively become very low cost multifunction cards combining payment and loyalty.

Allocating a slot to a particular retailer can be done automatically, when the card is first presented at that retailer's terminal. The retailer's RFM slot will remain active as long as the punch card program has not expired. After expiration, the slot becomes available once again. The customer 'chooses' which retailer slots are active simply by shopping regularly at those stores, and 'deactivates' slots by not shopping there anymore. The operation is completely automatic, for greater speed and minimum confusion.

It is always best not to encumber the flow of the transaction at the point of purchase by asking customers if they want to join the program. Retailers should simply be required to enter the amount of the transaction, just like they already do, leaving the software to perform complex marketing analysis on the customer's behaviour data in real-time, without requiring additional keystrokes or the use of complicated function keys.

The proliferation of add-on products for smart cards

Smart card loyalty systems require basic software for cards and terminals. In addition, other modules are often useful, like servers for card issuers and for chains of retailers, card readers for customers, Websites for customers to see where and how they can use their cards at local retailers.

As the capabilities of card systems increase and new features enter the mainstream, the market will require many more tools. Imagine the type of product retailers will need in order to manage their electronic coupons, or RFM at the item level. Such a product will certainly be useful to retailers, and even to consumer goods manufacturers who want to combine an electronic couponing campaign within a global marketing program around the launch of a new product. Imagine the type of product that allows a retailer to decide, in a very simple way, how to react to variations in historical RFM totals, how to know when a customer is losing interest, for example, and what to do. Imagine the type of product that allows a retailer to customize monthly mailings delivered to customers via e-mail. Ideas for add-on products are endless. The market for such products will be huge, since they will be of interest to virtually every retailer, and, for some products, will even be of interest to many cardholders as well.

Today, many card issuers provide a Website that their customers can use to obtain information about their credit card, pay bills and transfer money between accounts. As more and more PCs become equipped with smart card readers, these same Websites will provide additional capabilities. Customers will be able to insert their card and see a graphical representation of the electronic punch cards stored in the chip, presented exactly as they might appear if they were traditional paper punch cards, with holes punched on each of the visits that the customer has already made. Punch cards that are about to expire, or for which only one more visit is required to qualify for a free item or service, will appear at the top of the list.

Why doesn't my Palm Pilot have a smart card reader in it? I'd like to put my chip credit card in and see the status of my electronic punch cards and coupons. Why can't I get a coupon from the television during a commercial? Kinections Inc. is a company in California that is developing a device to do just that. It's a palm-sized 'smart wallet' that functions much like a television remote control, a two-way pager and a credit card all rolled into one. The device can retrieve and store electronic information and benefits directly from television, magazines and newspapers. Insert your card and icons concerning your electronic punch cards appear on the screen. Select the punch card you want to visualize and it appears on the screen exactly as it would appear on paper, punches and all.

Motorola and De La Rue, a major smart card manufacturer, are now integrating electronic commerce applications into portable cell phones. Once the basic platform is defined and becomes widely available, other features can be added, like the ability to receive coupons and special promotions. It will also be able to display the electronic punch cards a customer has on their smart credit card.

As a retailer, it would be fantastic to recognize a valuable customer at the point of purchase. It would be even better to recognize them as they walk into the store. Opening the door and entering a store already causes a bell to ring. What if the retailer were able to change the tone of the ring based on what type of customer walks in? Technically, this will soon be feasible. Regulatory requirements concerning privacy issues might make it less useful in some markets than in others.

Privacy issues melt away if the customer participates knowingly in the process. I want to be recognized as a valuable customer as much as possible, and be treated accordingly. Say a retailer tells me that once I've spent a total of $1000 in his store, I am automatically upgraded to gold status and receive many extra services. Essentially, the software allows a retailer to bump customers up in status by providing them with 'virtual gold cards', storing the information in the chip rather than having to physically change cards. When I stand in line at the delicatessen counter and take a number, it's always upsetting to be holding number 52 when the sign says '28'. What if once I reach gold customer status, the ticket machine recognizes me as a very valuable customer through an RF device that can read my card as it sits in my wallet? Then it could display '52' after customer 28 and before customer 29. Now there's an example of someone going out of their way to make things easier for their best customers.

All of these capabilities and more will be available thanks to innovations in smart cards, terminals, RF devices, barcode scanners and personal digital assistants. The convergence of payment methods, smart card technology and customer relationship marketing is becoming the driving force behind a whole new, immensely exciting industry.

Real-time customer recognition: three fundamental concepts

1. With mileage programs, customers acquire points across many different retailers which can then be used for payment. For retailers, this is just another way of giving a discount, but rather than providing it directly to the customer, the discount goes to the program operator who manages it for the customer, in exchange

for a commission. The system works when the program operator succeeds in adding significant value to the accumulated points and does not simply act as a financial clearing mechanism. Mileage programs must avoid being perceived by retailers and customers as a complicated deferred discount mechanism. With electronic punch cards, retailers and cardholders enter into a 'friction free' relationship in which the customer's loyalty is given in exchange for hefty discounts from the retailer without the need for a middleman to manage the discount, deferring it for future payment once the customer has acquired a sufficient number of points.

2 Each individual retailer's needs are different. Each store has its own distinct set of problems and opportunities. The individual store manager is usually the person best suited to responding quickly to local market requirements. Intelligence must be moved away from the centre and placed where it is most useful, directly at the point of purchase. Real-time technology provides customers with higher quality service and immediate access to benefits, without all the usual administrative overheads.

3 Web-based retailers offer real-time services because each customer's database is on-line all the time. Low cost multifunction smart credit/debit cards provide real-world retailers the same ability to efficiently access customer behaviour information in real time and offer instant services that can help them forge more intimate ties with their customers. A single file in the smart card can hold numerous electronic punch card programs.

4

ELECTRONIC COUPONS

❖

Here is a practice that fails ninety-eight percent of the time. There's nothing effective about an industry that fails ninety-eight percent of the time.

Procter & Gamble spokeswoman
talking about coupon redemption rates[14]

Consumer goods manufacturers and retailers both detest money-off coupons, but shoppers use and adore them. In 1996, Procter & Gamble began an eighteen-month test in upstate New York, eliminating coupons in favour of everyday low pricing. Many packaged goods manufacturers watched with great interest, some also began scaling down coupons, but consumer demand forced Procter & Gamble to call off the campaign four months ahead of schedule.

Manufacturers consider coupons an inefficient way to spend their marketing money. Redemption rates of less than 2 per cent of the coupons distributed in Sunday newspapers is a good example of the inefficiencies of the system.

Some retailers point out that although many customers want coupons to continue, there are an equal number of customers who see coupons as a hassle. They don't have the time to clip them, file them and manage them, yet they feel guilty that they're not taking advantage of discounts that they are entitled to.

Coupons tend not to accomplish what they set out to do – generate brand loyalty. People who use coupons usually buy whichever brand they have a coupon for. Procter & Gamble felt that lowering prices across the board may be a better way to lure customers and build loyalty. Studies have shown that the overall cost of distributing coupons through inefficient channels like the Sunday paper cost as much or more than the value of the discounts actually pocketed by shoppers. This suggests that a more efficient system doing away with coupons would allow customers to enjoy fully double the discount they currently experience with coupons, but without all the hassles.

Coupons are a century old tradition. Shoppers like them because they feel they are beating the system. This proved to be an insurmountable barrier for even industry giant Procter & Gamble.

Alternative delivery channels have met with some success. Companies such as Catalina Marketing Corporation have made inroads in targeted promotions, for example by delivering a coupon for nappies only to customers who have purchased other items indicating that they have a baby, like jars of Gerber's baby food. The redemption rate for Catalina's checkout coupons is 9 per cent, well above the 2 per cent for coupons delivered in Sunday newspapers.

Coupons appear to be here to stay. As new technologies develop, the goal of making coupons more efficient can be translated to the goal of achieving the highest redemption rates possible, the closer to 100 per cent the better. Ideally, new delivery methods should also integrate a loyalty mechanism, rewarding long-term frequent users better than opportunistic shoppers.

Using the smart card as an electronic coupon wallet

I can get punch cards from coffee shops, hairdressers and fast food restaurants. So why can't I also get a Procter & Gamble punch card for a free box of nappies after buying nine at full price? An electronic coupon can be loaded into my card for that free tenth box. Electronic couponing technology is expected to prove to be an important added value function capable of increasing card usage and retailer acceptance. E-coupons can be issued via an Internet Website, at kiosks or at in-store devices placed next to the product on promotion.

In 1993, the world's first electronic coupon wallet was tested in four rural supermarkets in France. I ran the small start-up company responsible for this initiative. The 'PromoCarte' smart card trial generated valuable knowledge that helped us better understand the interests of customers, retailers and packaged goods manufacturers. How did it work? The customer presented the smart card to the cashier, who inserted the card in the same payment terminal that handled bank credit and debit cards.

Scanned purchases were automatically compared to the coupons stored in the card's memory chip. When a couponed item was purchased, the system deducted the coupon from the customer's receipt and electronically sent the redeemed coupon to the processing centre. With every purchase, new coupons were added to the card, based on an analysis of scanned items and a second customer receipt was printed, listing all the coupons available on the customer's card.

A variety of program parameters were tried at the various store locations, such as free versus paid cards, or postal delivery of cards versus in-store delivery, among many others. The first test consisted of having customers fill out forms in the store to receive their card by post. A hefty 30 to 40 per cent of each store's customer base requested the card. One of the first discoveries was that when cards were personalized with the customer's name and sent by post, at no charge to the customer, up to 45 per cent of the cards never reached the customer's wallet and were simply never used. The 55 per cent that were used became active over a slow period of time, several weeks. But when cards were given out in the store, and sold at 10 francs a piece, the number of customers requesting the card dropped to between 20 and 25 per cent of the store's regular customers. However, 95 per cent of the cards immediately became active, as opposed to half of them never being used. By charging customers a small amount to participate, fewer cards were issued (25 per cent of the store's customers as opposed to 40 per cent) but more customers actually used them (25 per cent of all customers as opposed to 20 per cent). Active customers used their card on average 2.5 times per month.

Thirty consumer goods manufacturers participated in the pilot, representing leading name brands in 50 different product categories: Nestlé, Procter & Gamble, Quaker, Heineken and Lever Brothers, among many others. Redemption rates were very respectable, with an overall average of 20 per cent, almost 10 times that of coupons that you find in your Sunday paper. Coupons that were issued to current buyers of a specific brand enjoyed average rates of 44 per cent, some programs reaching over 70 per cent, while coupons issued to buyers of the competing brand were redeemed at the rate of 15 per cent.

Coupons were valid for a very short time, usually 30 to 45 days after being offered to the customer. The idea was to present an offer to a customer, make it very easy to use, remove barriers to the customer actually redeeming the coupon, then, if the coupon is not used quickly, make it disappear. Don't encumber customers with lots of stuff they don't want. As the number of messages addressed at customers continues to grow exponentially, people will come to differentiate between companies that throw the messages in their faces and those that politely suggest a message, quietly, discreetly, and that don't insist once the customer has indicated that no, a particular offer does not really interest them.

A big question prior to launching was whether or not customers would understand that coupons were electronically stored in the card's microprocessor chip. Late-night philosophical debates endlessly focused on this important issue. Pilot sites were deliberately chosen in smaller rural communities far from hi-tech Paris so that this question could be conclusively resolved.

To our great relief, the test proved that electronic couponing is indeed easily understood by customers, retailers and brand manufacturers. Customers quickly understood how the system worked and were sometimes annoyed when questioned about their level of comprehension during in-store surveys. Customers did keep their coupon list to use it on their next visit. Not only that, many immediately began writing their own shopping list on the back of the receipt. This was a completely unanticipated use of the receipt.

Just a few short years after smart cards had become a common part of French life, consumers nationwide clearly understood that information like coupons could be stored in their cards, updated and removed automatically through a point-of-sale terminal.

For retailers and consumer brand manufacturers the value proposition was simpler and pre-launch anxiety was less acute. Retailers benefit from a system financed by brand manufacturers, leaving margins intact. Brands enjoy an average redemption rate 10 times that of traditional paper coupons delivered through newspapers.

There was however one main drawback with the system, despite the fact that customer surveys showed a high level of satisfaction. A majority of cardholders stated that they used the card every time they went to the store, but the transaction data concerning those same customers showed differently. Customers who shopped four or five times per month were only generating 2.5 transactions. The card was apparently being presented only half the time. By watching customers as they went through the checkout lane, it was discovered that they simply neglected to present their card when they were not purchasing a couponed item. This behaviour, pushed to its extreme, quickly created a downward spiral. Every time a customer did not present their card, an opportunity was missed to issue new coupons based on purchases. And since new coupons were not added to the card, the customer had even less of a reason to pull out the card at the next visit.

On average, only 50 per cent of cards that had been used at least once remained active after two months. At that point, active cardholders redeemed an average of 1.5 coupons per month. These extremely poor results demonstrate that electronic coupons alone are an insufficient reason for customers to systematically use their card. What can be done to raise the number of coupons effectively redeemed? For one thing, making sure that customers present their card at each visit will ensure that every opportunity is taken to add new coupons to the card. The trial showed the need to include a loyalty mechanism for the retailer and, better yet, a payment mechanism as well.

Another parameter to consider is the number of coupons that are available on any given card at the same time. The PromoCarte was configured to contain a maximum of ten coupons at the same time. This number was chosen in order not to create a list too long to manage. A shopping list printed with more than ten items at three lines per item makes for a rather long list. In reality, active cards typically had three or four coupons at any one time, which was far too few. Customers did not have enough coupons to choose from. There was a critical threshold at three coupons. Cards that had only three coupons at a particular visit were very likely to disappear from the system, never to be seen again. It was estimated that the number of available coupons should be nine or ten. In order to double or triple the number of coupons any single customer has on their card, the total number of available product categories must be doubled or tripled. From 50 distinct product categories, the system should be increased to offer coupons from leading brands belonging to at least 100 distinct product categories, preferably 150.

This raises another issue. Leading brand manufacturers easily agree to participate in innovative programs that promise to optimize coupon promotion techniques. It wasn't too difficult to sign up brands in 50 main categories. Signing up an additional 100 categories is difficult work, but shouldn't be insurmountable. The test phase is easy and most product manufacturers routinely set aside marketing budgets to do small, localized tests. Manufacturers of course set aside much larger marketing budgets for their national promotional and advertising campaigns, which will in time cover large smart card-based electronic couponing programs. The problem is when the infrastructure has grown beyond a test phase but is not yet large enough to justify national marketing budgets. Consumer goods manufacturers generally don't have a marketing budget addressing this phase of development. At this critical period brand manufacturers can abandon en masse. At that point, costs of participating in an electronic couponing operation can no longer be seen as money invested to 'play around' with a new way to do coupons, but they cannot yet be seen as an investment in an alternative coupon delivery vehicle. The transition phase must be made to last a very short time.

Brand manufacturers themselves began providing hints on the type of product they would be willing to invest heavily in and support even during the critical transition period. One manufacturer after another kept telling us that electronic coupons are a good way of getting rid of paper coupons, but what would be really clever is some way of building loyalty, just like retailers do with their frequent buyer programs. What they were asking for, in fact, was their own electronic punch card program. A customer might participate in several punch card programs with leading consumer goods manufacturers like Procter & Gamble, Nestlé or Kraft. After spending say $200 on P&G products, the customer might get a $20 gift certificate, electronically loaded on the card. Today, there exists virtually no loyalty method that brand manufacturers can use that is practical and cost effective. So budgets will be allocated from a strategic perspective, as opposed to the tactical perspective that governs couponing budgets.

In retrospect, I am amazed that what brand manufacturers were asking for had not been addressed in the past. Not only that, the idea of applying volume purchasing rules to individual customers was so foreign that marketing managers were unable to formulate what they intuitively knew they needed. Taking the long view, once mass production had given way to national markets, these gave birth to mass merchandising, which gave birth to mass media. Our perceptions on how buyers and sellers want to work with each other have been warped ever since. Ultimately, the industrial age will be recognized as a relatively short period of time in which our initial relationship with the industrial process took us into social and economic systems foreign to our species. I am confident and optimistic that historians will one day show how we quickly absorbed the transition, eventually managing to establish a more natural relationship with technology.

A safe way to move into electronic couponing is to use a Trojan horse strategy. First issue smart payment cards that are adopted by customers and retailers for their electronic punch card capability. Once cards have been mass issued and customers have adopted electronic punch cards issued by retailers, you can begin introducing punch cards issued by brand manufacturers in limited regions. Start with a selection of programs in 100 to 150 product categories. Allow customers to select their P&G and Kraft punch cards on the Web or at kiosks or via special devices that can receive information from the television as customers watch an advertisement. That way they will be able to participate in the programs of their choice. Once the initial tests have been successful and the technology has been tweaked to function properly, launch instantly. Avoid spending any time at all in the grey area in between a pilot and a brand new nationally established coupon delivery method.

Electronic coupons: three fundamental concepts

1 Consumer goods manufacturers detest money-off coupons, which

are an inefficient way to spend their marketing money. Coupons tend not to accomplish what they set out to do – generate brand loyalty. People who use coupons usually buy whichever brand they have a coupon for.

2 As new technologies develop, the goal of making coupons more efficient can be translated to the goal of achieving the highest redemption rates possible, the closer to 100 per cent the better. Ideally, new delivery methods should also integrate a loyalty mechanism, rewarding long-term frequent users more than opportunistic shoppers.

3 Consumer goods manufacturers feel that electronic coupons are a good way of getting rid of paper coupons, but what they really want is some way of building loyalty. Today, there exists virtually no loyalty method that brand manufacturers can use that is practical and cost effective. Card issuers that offer manufacturers the ability to do their own electronic punch card programs will see main marketing budgets allocated from a strategic perspective, as opposed to the tactical perspective that governs couponing budgets.

5

MEASURING TRANSACTION RICHNESS

❖

The first time a checker at Vons wouldn't sell me grapes for 99 cents a pound without getting my Social Security number was the last time I shopped there. Ralphs is no better. Hughes is gone. Now I shop exclusively at Trader Joe's. There, the people recognize me and are friendly, whereas at the other stores only their big computers had any idea who I was.

Shopper in Pasadena, California[15]

Traditional loyalty methods using barcoded cards, social security numbers, or magnetic strip payment cards, are limited in their ability to address customers differently at the point-of-sale terminal. Their primary function is to collect customer data and upload it to a host. Any additional demands made on the system are very difficult to process.

As cards and terminals become more intelligent, payment transactions become far more information rich and can now respond differently to each customer, based not only on data concerning the customer's current transaction but also on data concerning many prior transactions. All of this information is available in the card, so the terminal can perform many more functions without slowing down the payment process.

The retailer's credit card point-of-sale terminal can now deliver not only coupons for free items or discounts, but also targeted information similar to direct marketing letters, customized according to each customer's actual behaviour. All of these features can be used independently of each other, or in unique combinations defined by the retailer. This all happens at the point-of-sale terminal, in real time.

Richer real-time transactions become a tool that helps store personnel better welcome their customers. In general, retailers and customers will both tend to prefer the payment method with the highest level of transaction richness.

Transaction Richness Quotient

Payment systems will differ widely in the level of richness they bring to each transaction. Magnetic strip systems have a low level of transaction richness, which is not to say that a central database management system cannot add richness to uploaded transactions. This indeed happens with data mining software. What we're talking about however, is real-time richness at the point-of-sale terminal, as the retailer is physically speaking with his or her customer. Smart card systems offer a potentially much higher level of transaction richness, as is intuitively obvious. It would be useful to be able to quantify the richness provided by each of the different payment methods.

The Transaction Richness Quotient is a measurement tool that can rigorously quantify the added value a particular payment method provides in real time at the moment of purchase. 'TRQ' is calculated by analysing the various parameters that are processed during the payment transaction. Parameters include information generated by the terminal – such as the purchase amount, the date and time of the transaction – as well as information provided by the customer's card, for example the cardholder ID, the date of this customer's prior transaction in this store, the number of transactions made during a specific period and the cumulative amount spent.

The objective of real-time marketing at the moment of purchase is of course to build customer loyalty, which is accomplished through three complementary objectives:

1. Customer knowledge generation – the ability to generate useful knowledge about the customer and make it instantly available at the payment terminal.
2. Reward attribution – the ability to attribute rewards to the customer immediately, such as discounts or free items.
3. Relationship building – the ability to establish a long-term relationship that is not necessarily based on financial rewards but

rather on 'soft' benefits like preferential treatment.

These three objectives should be developed concurrently in order to have a well balanced system. An unbalanced system may not have the desired impact on the customer's behaviour. For example, a system can go overboard on knowledge and neglect reward or relationship. This typically happens with data mining systems that analyse payment transaction data. The system itself generates knowledge, but all alone it is unable to have significant impact on how much the customer spends. The results of data mining systems must always feed other activities like database marketing in order to attribute rewards (by sending them to the customer) or build a long-term relationship. The Transaction Richness Quotient seeks to measure a particular system's ability to meet all three of these basic objectives concurrently, in real time, at the moment of purchase.

Each parameter that is processed at the payment transaction level will impact one or more of these objectives. A parameter that only impacts 'knowledge generation' will receive a score of 1. Another parameter that impacts 'knowledge generation' and 'reward attribution', for example, will receive a score of 2. Parameters that simultaneously impact 'knowledge', 'reward' and 'relationship' have a score of 3. After analysing all parameters, the scores are added up. This is the overall Transaction Richness Quotient.

Individual card implementations can be compared to one another in terms of each system's Transaction Richness Quotient. Subtleties like a particular system's enhanced capability in building relationships, as opposed to attributing financial rewards for example, become evident when one looks at the weight of each set of parameters within the overall Transaction Richness Quotient.

A large majority of loyalty programs are copies of airline frequent flier programs. Earn a mile or a point for every pound or dollar spent. Present your card at the checkout where your customer ID is linked to the transaction. You know how much you spent today and perhaps the receipt will tell you how many new points today's purchase entitles you to. That's about it. You might have accumulated a huge amount of points, or you might be in the store simply because the one you usually shop at is out of a particular product. The retailer has no way of knowing. Using systems with a low Transaction Richness Quotient is like navigating through clouds without a guidance system and without the ability to react in real time to avoid flying into a mountainside. Sure, traditional systems let you know lots of things about the customer once the data has been centralized and processed. But by the time the central system has come around to processing the data concerning a customer's specific visit, the customer has already left the store, gone home, worked a few weeks, and shopped over and over again since the date of the specific transaction that the central system is currently processing. The Transaction Richness Quotient is a way of measuring a system's ability to react in real time to a customer's actual behaviour.

The first smart card-based loyalty programs launched by large retailers like Shell and Boots in the UK, or by AOM French Airlines in France, were able to go a step further and increase the Transaction Richness Quotient by keeping the mileage account on the customer's smart card. This immediately provided improved visibility. The customer could now know, at each transaction, exactly how many miles had been accumulated. The miles are instantly available, so the day a Boots customer finally has enough to buy an expensive perfume or lotion, they don't need to call up Boots, write a letter, wait for a gift voucher, etc. They simply present their card along with their purchase and the points are debited from the chip.

The customer's perception of the quality of service provided by these new programs has soared. It's easy to know how many points you have accumulated since at every visit to Boots your receipt tells you. And because you know how many points you have, you don't have to call up a phone operator and ask for your balance. Unlike traditional loyalty programs, a smart loyalty card does not increase the number of customer service people required to answer account related questions. Another big benefit is that points can be redeemed by simply presenting the card at any point of sale and having the points debited from the chip. The customer does not have to request a gift voucher, and the retailer does not have to put in place a complicated paper voucher management system.

Improving loyalty programs by increasing their TRQ

The Boots, Shell and AOM programs are significantly better than programs with little or no real-time visibility, but they still don't give a clear enough view from a marketing standpoint. They still don't take into account a customer's actual behaviour. Proportionally speaking, all customers, whether they shop frequently or not, earn the same number of points. How can the Transaction Richness Quotient be increased with these programs? If Boots were to process recency, frequency and monetary value parameters at the chip level, customers can be rewarded based on tiers. The first £50 spent in a given period might earn 2 per cent in points, the next £50 might earn 3 per cent, while anything over £100 might earn 4 per cent. This would result in a significant increase in transaction richness. In addition, by giving less to casual customers and more to frequent customers, the overall program costs far less than giving everyone the same 3 per cent.

Shell's Smart program is a multi-retailer program that allows customers to earn points at a growing number of retail chains. Again, as in the Boots program, points are given based on a simple purchase amount. Using the electronic punch card concept, an individual retailer participating in Smart can vary the points based on usage of the card at that particular shop. Say, by offering 2 per cent in points to anyone who walks in the door carrying a Shell Smart card, and by offering an additional 3 per cent to customers who spend a minimum of £100 in a single month.

Many franchise chains will find that electronic punch cards allow them to create a chain-wide loyalty program that finally satisfies individual store owners. With many franchises, the store owner feels he's building loyalty to the chain, as opposed to his particular store. If the loyalty program and the rewards are completely paid by central marketing, that's all right. The budget is coming out of whatever each franchisee pays in marketing fees to the franchiser. When a franchise chain's central marketing tries to create a loyalty program in which rewards are financed by individual stores, the whole project usually falls apart. 'Why should I give a free Big Mac to someone who earned his loyalty points somewhere else?', asks a store manager. 'I want to give a Big Mac to customers who earned their points in *my* restaurant.'

This is a common complaint. Retailers who participate in multi-store or multi-chain programs want to give loyalty points to frequent customers, but they want those points to be valid only in their store.

Electronic punch cards solve this problem. Central marketing can offer a loyalty card program that allows each store to run its own electronic punch card. As a customer, you receive your card initially empty. When you go to a store that accepts the card, that store's electronic punch card is automatically added to the chip. The receipt tells you that after three more visits to this restaurant, for example, you will receive a free Big Mac. When you go to other participating stores, their punch cards will also be added. This allows each retailer to reward customers loyal to that individual store.

Financial institutions involved in private and co-branded card programs know that in the US and elsewhere, the market for traditional 'frequent flier mileage' type programs is saturated. Today, smart cards using electronic punch card technology can open new markets for numerous franchise chains that could not adapt mileage programs to function in their environment.

Measuring the richness of payment transactions

Many e-purse pilots have been launched to compete with cash and cheques. These transactions can also be measured. The Transaction Richness Quotient for cash appears intuitively to be very close to zero, in any case well below cheques and e-purse. Cheques can be used to generate knowledge about the customer. Many retailers collect their customers' names and addresses from cheques, but with cash, a retailer doesn't know who the customer is. So the Transaction Richness Quotient for cheques would appear well above that of cash. In reality, it's just the opposite in many cases. How can cash sometimes have a higher Transaction Richness Quotient than cheques? Cheques can generate knowledge, but cash can often generate rewards to the customer. In many places things cost less when paid in cash. Electricians, plumbers and gardeners in the south of France can routinely quote you two prices, the normal price and the cash price (if you want to know more about similar idiosyncrasies of southern French culture, see Peter Mayle's book *A Year in Provence* – although it reads like fiction, much of it is true). Cash avoids VAT, income and professional taxes. The savings can be shared with the customer.

What is the Transaction Richness Quotient of an e-purse? It is about the same as cash, unless cash provides a discount reward that e-purse does not, in which case cash may in fact have a higher TRQ than the e-purse.

Customers will tend to prefer the payment method with the highest Transaction Richness Quotient. If you are launching a debit card or an e-purse in a specific environment where cash has a particularly high Transaction Richness Quotient, you must take extra care to make your card product very attractive.

Magnetic strip credit card systems generally have a Transaction Richness Quotient of around 2. Electronic cash implementations like Visa Cash, Mondex or Proton have a quotient of around 5, comparable to France's chip-based debit card system. An early implementation of Welcome Real-time's loyalty software for smart cards had a Transaction Richness Quotient of 23. That was four years ago. The most recent release has gone just above 100. At the end of 2000, the TRQ is expected to reach 170.

The Transaction Richness Quotient is increasing rapidly

Around fifteen years ago, magnetic strip payment cards were everywhere. The financial industry then began moving to smart cards, initially in France for debit cards, then in other parts of Europe for electronic cash. Retailers soon began using smart cards to implement loyalty point mileage systems. More recently, companies have begun deploying electronic punch card functionality.

The Transaction Richness Quotient has quietly been doubling approximately every eighteen months. Meanwhile, transaction times have been cut dramatically; from up to one minute with magnetic strip credit cards (for an on-line authorization and manual ID and signature verification), transactions have come down to well under 10 seconds with smart cards. Something quite similar to Moore's law seems to be applicable here. Once you think about it, there's no reason why the smart card industry should not be governed by principles similar to those that drive the microprocessor industry. In light of the huge potential for new, complex features that smart card software will provide in the future, the Transaction Richness Quotient will continue to increase over the next few years. Why wouldn't it follow general computing trends and simply double every eighteen months?

The impact on card and terminal hardware

The Transaction Richness Quotient is a standard way of measuring the richness of a particular payment method. It does not measure how the payment transaction physically happens, for example, how long it takes or how much investment is required in infrastructure costs.

Cash is fast and requires no investment in infrastructure (the cost of handling cash is an ongoing expense that most retailers don't quantify). Credit cards processed on-line for authorizations are slow and require an expensive infrastructure.

Individual card technologies, terminals, and service providers can have an impact on the speed of transactions and their cost. Calculating Transaction Richness Quotient ratios per second and per dollar for each individual combination of infrastructure elements can help in choosing the right suppliers. A specific terminal model that offers fast transaction processing and printing will have a higher TRQ per second ratio than other terminals. This is an important ratio because point-of-sale payment transactions, no matter how rich, must be carried out in less than a handful of seconds.

Measuring Transaction Richness: three fundamental concepts

1. Retailers and customers both tend to prefer the payment method with the highest level of transaction richness. Richness is measured by the Transaction Richness Quotient (TRQ), which seeks to measure a particular system's ability to concurrently generate knowledge, attribute rewards and build relationships with customers in real time, at the moment of purchase.
2. Smart card systems with a high TRQ allow retailers to adapt

rewards based on actual customer value, giving proportionally less to casual customers and more to frequent customers; and by taking into account the customer's exact relationship with that particular retailer.

3 The Transaction Richness Quotient has been doubling approximately every eighteen months. Magnetic strip credit card systems generally have a TRQ of around 2. Electronic cash has a quotient of around 5. Electronic punch card systems are currently at 100 and will soon reach 170. The Transaction Richness Quotient will continue to increase rapidly over the next few years, following general computer industry trends.

6

LOYALTY BRANDS

❖

Then wear the gold hat, if that will move her;
If you can bounce high, bounce for her too,
Till she cry, 'Lover, gold-hatted, high-bouncing
lover, I must have you'.

Thomas Parke d'Invilliers[16]

The little rectangles of plastic that we use to pay with would all be identical if it weren't for each card's distinctive brand along with the services that the brand delivers and the promise that the brand represents. When customers decide which cards earn the privilege of residing in their wallets, they don't do it by comparing quality and thickness of plastics, nor the size of the chip or the microprocessor's clock speed. They decide based on the services and features offered by a particular brand.

When the brand represents services and features that the customer finds useful, the card has a good chance of finding a place in the customer's wallet. Once the card is in the customer's wallet, half the battle is won. But only half. In order to completely succeed, the card must be made irresistible. If the brand is irresistible, the card receives preferential treatment by both customers and retailers. The goal of course is to do everything possible so that customers and merchants feel that they 'must have that card'.

When a retailer issues a private card, the only brand that appears is the retailer's. The card can usually be used only at that retailer's chain, although in some cases the card is also honoured at other stores that have entered into a partnership with the primary retailer. Private cards work best when the retailer has such a compelling market presence and product offering that customers feel they simply cannot do without the card. For the vast majority of retailers, this is very hard to do. That's why private cards tend to be of interest to a very small number of already loyal customers.

With a co-branded card, two brands appear on the front, the retailer's brand (for example, American Airlines) and a payment brand (such as Visa). Because co-branded cards can be used anywhere the card's payment brand is accepted, these cards are easier to launch than private cards. The issuing bank's brand may also appear on the front of the card (like Citibank), otherwise it is placed on the back. A new trend that may begin to appear is to place the payment brand on the back of the card. Co-branded cards generally attract a larger share of a retailer's customers than private cards. It is generally easier to place a co-branded card into a customer's wallet than a private card. However, making the card irresistible remains just as difficult.

Branding is a simple way to quickly tell customers and retailers how the card works. The various brands placed on the card help to clarify a potentially confusing product. In the Citibank/Visa/American Airlines example, customers understand that they can use the card for payment everywhere Visa is accepted. They also understand that they get special rewards with American Airlines. If they want to increase their credit limit or have questions concerning their account, they know they need to call Citibank.

Smart cards create new branding opportunities

The market for private cards and co-branded cards is already very mature. Branding issues have long been resolved. What about the completely new market of running electronic punch card programs on a chip-based credit card? Just as prior card products required specific branding strategies adapted to specific markets, the new smart card market will require its own distinct branding strategy.

Most retailers that offer paper punch card programs today have no intention of offering private or co-branded payment cards. If they did, they would have done so long ago. These retailers need to be able to link their electronic punch cards to payment products issued by others. Ideally, they should be able to address the largest possible customer base.

When a retailer runs a loyalty program on the back of someone else's smart card, alongside potentially dozens of other retailers, no single merchant brand will appear on the card. There is potentially a great deal of confusion if branding issues are not properly addressed.

How do cardholders know which retailers offer rewards? How do they know whether they get a free sandwich at their fourth visit or after spending £10? How do retailers know which cards they honour in their electronic punch card program? They can't simply say 'Get a free sandwich the fourth time you pay with your chip card' because not all chip cards will have the compatible loyalty software loaded. Retailers might be able to specifically identify a bank, as in 'Get a free sandwich the fourth time you pay with your new ABC Bank smart card'. One drawback to this is that it is cumbersome, especially if ABC Bank has other smart cards that would not qualify, say ABC employees' cards used to pay at company vending machines and photocopiers. Another drawback is that it potentially limits the retailer's program to a very small proportion of customers.

A loyalty brand is required, something like 'Instant Rewards'. That way, the cardholder and the retailer can refer to the program quickly: 'Get a free sandwich every fourth time you pay with your Instant Rewards card'.

If your card program does not represent a sufficient percentage of a retailer's customer base, you can allow the retailer to supply single-function cards, loyalty without payment, also carrying your 'Instant Rewards' brand. You can also license your brand to other card issuers such as transport authorities, phone companies and financial institutions. If you are a bank that issues Visa or MasterCard credit cards, licensing your loyalty brand to competitors might sound like a strange idea, but if you own a dominant brand, you can control how it is used by the competition. For example, you may decide to allow competing banks to issue dual-function payment/loyalty cards and insist that single-function loyalty cards only be issued by your company. That way, you can retain an ongoing relationship with customers who will at some time or another become customers for your credit card products.

Once a retailer places your 'Instant Rewards' loyalty brand in the store and begins running an electronic punch card program linked to that brand, you have a virtually permanent relationship with that retailer. It becomes very difficult to back out. As other retailers join the program, you begin to build momentum. Customers prefer your brand because more retailers use it for their programs. Retailers prefer your brand because more of their customers are cardholders.

Using loyalty brands in conjunction with private and co-branded cards

A loyalty brand like 'Instant Rewards' can coexist with private cards and co-branded cards issued by merchants. Take a fast food chain, for example, Pizza Hut. The 'Instant Rewards' brand manager can offer several options to Pizza Hut:

- Pizza Hut can offer a free pizza the fifth time a customer pays with their 'Instant Rewards' payment card.
- Pizza Hut can encourage customers to sign up for an 'Instant Rewards' credit card.
- Pizza Hut can provide 'Instant Rewards' loyalty cards to customers who don't have an 'Instant Rewards' payment card issued by their bank. Many retailers will be concerned about potentially alienating good customers that don't bank with an 'Instant Rewards' issuer, or who simply don't want another credit card.
- Pizza Hut can provide loyalty cards with their own brand alongside 'Instant Rewards'.

Some large retailers will insist on issuing their own cards with no other brand on them. This is the model that the market is familiar with, so the reaction is normal. Even if a retailer decides today to do their own, exclusive, private program on their own cards, they will one day nevertheless want to include customers that participate in the potentially much larger 'Instant Rewards' program. This will give them access to a vast customer base of people whose cards are paid for by someone else. Once a retailer understands that they can run their own private, electronic punch card program on someone else's card, and that they don't have to purchase cards themselves, maintain them and keep them alive and active, the overall business case becomes clear. Cost savings are huge. It is precisely the high cost of properly running a card program that has prevented most retailers from issuing a private loyalty card.

Once an 'Instant Rewards' type program takes off, retailers will have no reason not to become involved in it. They might not see this coming today, but the loyalty brand manager should plan for it now and present retailers with clear, attractive long-term options.

Who will create multi-issuer loyalty brands?

The banking industry, led by Visa and MasterCard, immediately comes to mind. Will the associations create these brands, or will it be some of the larger banks themselves, who then go on to license their brand to smaller banks?

Retailer acquirers have a particularly good reason to become loyalty brand managers. Today, acquirers recruit retailers by negotiating transaction fees. Differentiation based primarily on pricing has created an industry with razor-thin margins. If an acquirer creates a loyalty brand, launches it and manages to license the brand to several card issuers, retailer recruitment can become significantly easier. Existing retailer contracts can be made more secure. Negotiations need no longer be based only on pricing, since a whole new discussion on the added value of the loyalty brand suddenly becomes possible. The acquirer can more easily win over new retailers for payment transaction processing when the loyalty brand is part of the overall service package offered to retailers.

Once a retailer switches over to a new payment acquirer in order to link their loyalty program to the acquirer's loyalty brand, the retailer becomes dependent on the brand owner for a long period. As time goes by, it becomes more and more difficult for the retailer to stop using the brand and hence to switch again to a rival acquirer.

Large banks that are both card issuers and retailer acquirers will be able to enjoy combined benefits from their loyalty brand. In addition to new acquirer revenues, these organizations can enjoy a higher share of the retailer's transactions and reduced cardholder acquisition costs. Participating retailers will want to encourage their customers to become 'Instant Rewards' cardholders and will gladly make cards directly available at the point of sale. Active merchant participation in recruiting new cardholders is far more effective and less expensive than mass mailings offering enticements and introductory offers of free credit.

Telephone companies should be able to leverage a multi-issuer loyalty brand into a powerful development strategy. They tend to blanket a region with telephone cards already. If these phone cards are loyalty based they can represent a high portion of a retailer's customer base. The telephone company's loyalty brand can then be licensed quite easily to financial institutions.

What about marketing coalitions like Air Miles? Many of these are already multi-issuer loyalty brand managers and will be able to expand their current services with new functionality based on the smart card's capabilities.

All of these organizations will launch and experiment with multi-issuer loyalty brands over the near term.

What do companies stand to lose by not owning a dominant multi-issuer loyalty brand? Quite simply a very privileged relationship with retailers and cardholders. This is a key element of the smart card's overall value proposition. Control of a smart card loyalty brand will prove to be a very valuable asset.

Emerging market expansion strategies

From a financial institution's perspective, private store cards and co-branded cards are zero-sum games. If I win a deal to create a co-branded card for a particular retail chain, my competitors lose. If they win the deal,

I lose. The same is true for general purpose credit cards. If I can persuade a customer to use my bank's credit card more often, that customer will use other bank cards less often. I win, you lose. You win, I lose.

This was not always the case. Zero-sum games are a characteristic of a mature market. The first companies to issue credit cards started by taking big chunks of market share away from cash and cheques. It was only when the market began approaching a point of saturation that credit card companies were forced to battle to take market share away from each other.

Long-term growth in all markets is initially powered by relatively short bursts of non-zero sum competition, followed by relatively long periods in which the market becomes increasingly mature. The market structure is defined at the very early stages. Initial market leaders tend to maintain their position of strength throughout the life of the market, until a new paradigm comes along and makes the old approach obsolete. Competition becomes increasingly more difficult as the market gradually becomes a zero-sum game.

Today, smart credit and debit cards are particularly effective as a means of taking further market share away from cash and cheques. The opportunity again exists to take advantage of the upcoming burst of growth that will inevitably happen as several financial institutions stake their positions as leading providers of card products in this new market.

A multi-issuer loyalty brand like 'Instant Rewards' is the single most important key to triggering a snowballing effect that will attract retailers, customers and card issuers. As the web of interrelations is woven tighter, its strength grows exponentially. This is the law of increasing returns. As more retailers join, cardholders find the program more compelling. As more cardholders join and become active participants, retailers find the program irresistible. As card issuers begin adopting the brand on their cards, other card issuers become attracted to the program and find they must participate *now* if they are to be part of the entire market's burst of growth.

In zero-sum games you always try to hide your strategy. But in non-zero sum games you might want to announce your strategy in public so that the other players need to adapt to it. When Gorbachev did away with 10 000 tanks, he announced it clearly and unilaterally put an end to the cold war by eliminating the West's justification to continue investing heavily in weaponry. Being secretive would have had no effect.

Emerging market expansion strategies are based on co-evolution and non-zero sum games. During this phase of development, control and secrecy are counterproductive. This is why it is wise to openly license your loyalty brand to many other issuers, even if they are your competitors. This will help establish your company as the market leader for a long period of time.

Kevin Coyne, of McKinsey & Co., explored the competitive dynamics of network-based businesses in an article published in the *Harvard Business Review*.[17] He looked specifically at how retail banking established an ATM network in the US. In the late 1970s, Citibank became the first to offer large numbers of ATMs at its branches in New York, providing brand new 24-hour service to customers. Other banks had installed a few ATMs as well, but none was widespread. Citibank's proprietary network gave it an enormous advantage. In response, rivals, led by Chemical Bank, banded together and formed the Plus network. In network-based businesses, smaller players generally succeed when they band together to compete against the dominant provider. Citibank initially declined to join Plus, but as often happens in network wars, the greater combined numbers of the small players overwhelmed the large single player. Although Citibank was the most convenient single bank, it could not match the combined presence of the other banks. That would have required placing a proprietary outlet almost everywhere anyone wanted to do banking. The Plus network was the winner in the battle for perceived ubiquity.

Dee Hock, founder of Visa International, recalls the early days of the credit card industry when banks failed at establishing proprietary retailer acceptance networks.[18]

> One bank issued cards with a hole in the center and supplied merchants with imprinters which had a matching steel peg in the bed to shatter the cards of competitors. No stupid sharing of point-of-sale devices for them. Let the merchant worry about irate customers. Another bank, dreaming of riches from rental fees, methodically removed competitor's imprinters and installed their own, only to be thrown in the slammer by the local sheriff, who happened to be a loyal customer of the competitor.

Here's another example of a failed attempt at creating a proprietary network.

> Are you familiar with the joint venture of Bank of America and American Express, who proposed to make tenant farmers of the banking community by creating an authorization system capitalized by the users of the system, but owned solely by the two of them? Or how it collapsed when an industry-wide committee proposed building a system to be jointly owned by all banks, travel card companies and retailers? Or how that, in turn, collapsed when Visa withdrew to form its own system? On dozens of such fragile, seemingly innocuous hinges, did the industry turn. How radically different would it be had it been led in another direction by the managers of the time? How many seemingly innocuous decisions are now being made which will shape the industry for decades to come? How perceptive are you about those decisions? How capable of influencing them?

Interesting questions indeed.

Creating a network is expensive and risky if it is done in a closed, proprietary manner. Developing a network becomes far less expensive and virtually risk free if it is shared among competitors. This suggests that the strategy for a strong, established market leader would be to launch a proprietary network to gain time-to-market advantages, then suddenly open the network up to competitors once they appear to be on the verge of banding together.

Swarms, languages, network externalities and the law of increasing returns

Have you ever seen a swarm of bees take off? Kevin Kelly provides a lovely description of the process in his book *Out of Control*. When bees outgrow their hives and feel the need to move to a new place to live, individuals will search out potential sites. A bee will fly back to the hive and dance to communicate what it has just found, 'Go over there, it's really nice'. Other bees watch, then go out to see for themselves. They come back and say something like, 'Yeah, it really is nice'. The intensity of the dance, and the number of bees that concur, encourage more and more bees to go and check out the new place. Multiple sites are championed in this fashion, but eventually critical mass builds for one particular site, the hive becomes more and more excited, finally it bursts out, rises into the air, hovering over the empty hive, like a soul leaving its body, then heads out toward the new site. This is pure democracy in action. Each bee votes, and the majority wins.

Languages evolve in a similar manner. Speaking a language shared by many people is more valuable than understanding and speaking a language limited to a few individuals in a remote mountain village. Looking at language evolution and changing cultural influences across continents over several centuries, one can clearly see things that feel similar to the swarming effect that happens when bees move to a new site. A language becomes more and more valuable as more and more people speak it. At one point it achieves critical mass and then increases exponentially, filling up an entire region, country or continent. Once critical mass is achieved, things happen very quickly. Two generations is enough to completely switch to the dominating language.

Swarms and languages illustrate what economists call 'network externalities'. If the value of an object increases as more and more people use it, network externalities happen. A single fax machine has no value. Two fax machines make each machine twice as valuable. The size of the network counts far more than the machine itself. Telephones benefit from the same effect. So does the Internet. A network's value grows even faster than the number of members added to it. A 10 per cent increase in customers for a company that does not benefit from increasing returns may increase its revenue by 10 per cent. But a 10 per cent increase in customers for a telephone company could raise revenues by 20 per cent because of the exponentially greater numbers of connections between each member of the network. Most software packages also benefit from this, like operating systems of course, but also word-processing applications, spreadsheets and presentation managers. If people around you all use Word, you are better off also using Word so that you can easily share files, learning and advice.

Multi-issuer loyalty brands will also react to network externalities. The market will prefer an 'Instant Rewards' type brand that many customers have on their cards, and that many retailers use for their electronic punch card programs. As more and more cards carry a particular brand, more and more retailers will want to accept the brand. This in turn makes the brand more valuable to other customers that don't yet have an 'Instant Rewards' branded card. They become cardholders. That makes the whole network more valuable to other retailers. They join in. And so on and so forth. Once critical mass is reached things accelerate. The 'Instant Rewards' network grows exponentially in value and soon becomes irresistible to customers and retailers who appear to be saying, 'Card, I must have you!'

Markets that benefit from network externalities abhor niche strategies. In the seventies and eighties, a number of industries developed specialized proprietary networks linking buyers and sellers together for order entry, invoicing and inventory management. Various incompatible EDI (Electronic Data Interchange) protocols were defined and put into place at great cost to their champions. Today, most of these have all but disappeared, swallowed up by the mass market Internet. The remaining EDI companies were able to understand that their intellectual property was not in the area of infrastructure like communications protocols, but in content which is easily transferred to the Internet. Of course, the Internet itself evolved out of a niche market of scientists and university researchers. The point is that with network externalities, the niche phase is never long lasting. It will always be replaced by a mass deployment phase at some point. Niche markets will always be better addressed by an inexpensive product offering aimed at the mass market.

At the beginning, niche brands for electronic punch cards may very well emerge in some markets. Brands may appear for specific cities or regions, something like 'New York Rewards', with an apple as part of its logo. Other brands may appear for categories of customers, like France's youth card, 'Carte Jeunes', a smart card brand dedicated to the under-26 crowd. However, because of network externalities, the market will always prefer, if not demand, a single, widely available electronic punch card brand. A common brand will be made available to a much larger number of cardholders, which will attract a much larger number of retailers. It is inevitable. Just as the Internet instantly swallowed up professional EDI networks the moment it entered the mass market scene, a common multi-issuer electronic punch card brand will prove to be irresistible. This is also why payment brands like Visa and MasterCard are so powerful. It is unthinkable for retailers to have to enter into agreements with hundreds of individual banks in order to process credit cards that only carry the bank's name. Electronic punch cards require the same strategy. Because retailers don't sign payment processing agreements with numerous banks (and don't paste hundreds of bank logos on their doors) they will not sign loyalty processing agreements with numerous individual banks.

It is virtually impossible to fight against network externalities or the law of increasing returns, short of persuading governments and cartels to restrict the use of payment cards as marketing vehicles (which is the case in France today) or to simply outlaw the practice of loyalty marketing in general (as is the case in Germany). Early champions of niche products and services often spend huge fortunes fighting a losing battle against the law of increasing returns. Early EDI protagonists never recuperated their investments in proprietary protocols that were easily overrun by the Internet. They were forced to transfer their services to the Web.

Companies that battle against the law of increasing returns lose a great deal of money. Companies that encourage network externalities and help a fledgling product category emerge and begin benefiting from the law of increasing returns almost always end up in a dominant market position for many years, at least throughout the lifecycle of that product category.

This represents a tremendous opportunity for companies that are now beginning to champion a common, multi-issuer electronic punch card brand.

Entry barriers

For all the benefits, opportunities and competitive positioning that smart card loyalty provides, there exists an equal amount of risk if one does not reflect sufficiently on how the brand is managed. Smart card loyalty and electronic punch cards are a completely new business activity. Risks are not yet necessarily understood, even by people who are beginning to supply the system's underlying infrastructure technology. Launching a multi-issuer loyalty brand presents a number of complex risks that should be understood at the beginning of the project. Many companies will learn the hard way, I know we did.

When launching a multi-issuer loyalty brand, here are some of the issues that we found must be addressed in great detail prior to launch and that must be intimately understood in terms of potential future impact:

- What is the brand's promise? This should be evident in the brand's name. Visa founder, Dee Hock, says it took two years to come up with the name Visa.[19] He says what you need is a word that people anywhere in the world will understand and can communicate.
- Does the brand denote a loyalty program, like Air Miles, Membership Rewards or Smart? Or does the brand denote a function, method or technology, like Instant Rewards? Companies that are unable to grasp the distinction will most likely fail. Loyalty program brands are best adapted to alliances of complementary retailers, while loyalty function brands are best adapted to large, widespread deployment of cards allowing individual retailers to perform loyalty marketing tailored to their stores. Think content versus conduit.
- How can you reach critical mass instantly, upon launch? Assume you have established a simple, clear and easy to understand value proposition for all parties. Good. You have a potential upward spiral ready to take off. But how do you make it take off? How do you light a fire under it? Where is the match? The risk here is to spend several years languishing in pilots without ever actually reaching critical mass.
- Some of the most evident risks with new technology are of course related to intellectual property. Make sure the system you plan to use has no outstanding or potential patent issues limiting your ability to deploy. Don't wait to have millions of cards issued before discovering a patent infringement problem that may require you to replace those cards. Patents are a manageable risk. There are enough lawyers around that understand the details of intellectual property protection.

Marketing agencies can help you work through some of these issues, but they're learning too. We have spent a large amount of money with marketing agencies, but found that we were mostly paying them to learn. We had no desire to spend a fortune teaching an entire industry before receiving anything in return, so we hired our own senior marketing executives and offered stock. This allowed us to think through the complexities of this new industry and develop robust strategies based on our technology without losing time.

Complexity is a potent entry barrier. Once you have grasped the details, you're off. The first to understand and see which way to turn invariably wins and remains ahead for a long period of time.

There is an analogy in the birth of the credit card industry itself. Dee Hock describes Visa's origins in the late 1960s, when the credit card industry was on the brink of disaster. The Visa card had its genesis a decade earlier as a California service of the Bank of America called BankAmericard. Concerned with possible erosion of their customers, five California banks jointly launched MasterCharge in 1966. In turn, Bank of America franchised its service. Other large banks quickly launched proprietary cards and offered franchises. Action and reaction were soon rampant. Bank after bank mass issued cards with little regard for customer qualifications, while television screamed such blather as, 'The card you won't go berserk with', a challenge the public accepted with enthusiasm.

'By 1968 the infant industry was out of control,' says Dee Hock.[20]

> Operating, credit and fraud losses were believed to be in the tens of millions of dollars. *Life* magazine ran a cover story depicting banks as Icarus flying to the sun on wings of plastic, beneath was a red sea labeled losses, into which banks were soon to plunge, wings melted, and drown. In the midst of the mess, Bank of America called a meeting of its licensees to discuss operating problems. It quickly disintegrated into acrimonious argument. In desperation, the bank proposed forming a committee of seven, of which I was one, to propose solutions to the more critical problems, which the bank would then attempt to impose.
>
> Within six months, a complex of regional and national committees had been formed, which had but one redeeming quality: it allowed organized information about problems to emerge. They were much worse than anyone had imagined, far beyond possibility of correction by the existing organization. Losses were not in the tens of millions, but in the hundreds of millions, and accelerating.
>
> And suddenly, like a diamond in the dirt, there it lay. The need for a new organization and a precarious toehold from which to make the attempt.

Visa was established according to rules that on the one hand ensured proper cooperation among banks for core issues that were needed for the system to work, and on the other hand allowed banks to continue competing against one another. The organization was designed to be highly decentralized and highly collaborative. Everything is pushed out to the periphery of the organization, to the member banks themselves. Bylaws encourage members to compete and innovate as much as possible, while at the same time, in a narrow band of activity essential to the success of the whole, members engage in the most intense cooperation.

The last fifty years and the next five

By understanding how payment cards have evolved over the past fifty years, we can begin to see where the market is going over the next five years. It transpires that a common loyalty brand is just a new twist on how the payment card industry has addressed other issues in the past.

Store credit cards were born in the 1940s. These cards were accepted for payment at a single large department store or chain and were issued by a single financial institution. Sears, JC Penney and Kmart are examples of organizations that have issued private cards.

The first store cards quickly gave rise to a number of local credit cards issued by large regional banks and accepted across a variety of merchants. In the late 1940s, a number of US banks started issuing their customers scrip that could be used like cash in local shops. The practice was formalized in 1951, when the Franklin National Bank introduced the first modern credit card.

In the 1960s, local credit card issuers joined forces, giving rise to the associations that later became known as Visa and MasterCard. Bank of America was first to extend the credit card throughout the United States by introducing the BankAmericard, which later became Visa, and by franchising a single bank in each large city. These banks recruited local merchants and enrolled cardholders.

At around the same time, loyalty programs began proliferating. Punch cards are valid at a single retailer location or chain and are not linked to any particular payment method. Co-branded cards issued by a single issuer are valid at many retailer locations for payment, but only provide loyalty rewards at a single retailer location or chain. Private loyalty cards are simply store credit cards with a loyalty program attached.

Then loyalty programs became more complex. Single issuers like American Express developed programs like Membership Rewards, pooling mileage points from many retailers into a common bucket. In some cases, the same reward program was adopted by multiple-card issuers, giving rise to multi-issuer loyalty brands like Air Miles (initially launched by British Airways) and UK's Smart program (initially launched by Shell).

Let's look at what has been happening here. During the market's early phase, we can identify two main product evolutions over a fifteen-year period (store credit cards and local credit cards), and four main product evolutions over the next ten-year period (card associations, store loyalty cards, co-branded cards and mileage programs). The trend is accelerating and there is no reason to believe it will not continue.

Smart cards provide an opportunity to create more powerful card products. By applying the market rule of concurrently addressing the needs of banks, retailers and customers, and by simply extrapolating what has been happening in the payment card industry over the last fifty years, we can develop a vision of where the market is heading.

We're already beginning to see the first single-issuer smart cards that let multiple retailers each run their own electronic punch card program on the same smart payment card. This is a recombination of simple credit cards (like those issued prior to the existence of Visa and MasterCard), with punch cards that already existed long before credit cards came out. As such, it represents an evolution that uses new technology to create new combinations of proven products and services, each with their own established business model.

The next step will be the creation of loyalty brands used by multiple-card issuers which allow retailers to run their electronic punch card programs across a large number of cards. Given the payment industry's long-standing interest in uniting around a few common payment brands, the industry will soon be moving toward a few common smart card loyalty brands.

Loyalty brands: three fundamental concepts

1. Customers decide which cards earn the privilege of residing in their wallets based on the services and features offered by a particular brand, not the card's technology. The market for private cards and co-branded cards is already very mature. Branding issues have long been resolved. The new smart card market, based on electronic punch card technology, will require its own distinct branding strategy. A loyalty brand is required, something like 'Instant Rewards' so cardholders and retailers easily know where the cards generate rewards and which customers are entitled to them.
2. Once a retailer places your 'Instant Rewards' loyalty brand in the store and begins running an electronic punch card program linked to that brand, you have a virtually permanent relationship with that retailer. It becomes very difficult for the retailer to switch to another provider. As more retailers participate, cardholders find the program more compelling. As more cardholders join and become active participants, retailers find the program irresistible. As card issuers begin adopting the brand on their cards, other card issuers become attracted to the program and find they must participate *now* if they are to be part of the entire market's burst of growth.
3. Initial market leaders tend to maintain their position of strength throughout the life of the market, until a new paradigm comes along and makes the old approach obsolete. The opportunity again exists to take advantage of the upcoming burst of growth as the payment card market expands into new merchant sectors that do not yet accept plastic cards. Emerging market expansion strategies are based on co-evolution and non-zero sum games. During this phase of development it is wise to openly license your loyalty brand to many other issuers. This will help establish your company as the market leader for a long period of time.

7

EVOLUTIONS IN SMART CARD AND TERMINAL PLATFORMS

❖

IBM's grip on the IT world was loosened by the growth of the personal computer which it created, and the growing influence of Microsoft, which it nourished. But within five years Microsoft's influence overtook IBM. Similarly, card manufacturers in today's smart card market have passed control over to the independent software vendors. For implementers and adopters, the strategic decision now lies in selecting and developing card applications, not in knowing which card technology to deploy.

Duncan Brown, senior analyst at Ovum[21]

Hundreds of thousands of terminals must be replaced or upgraded to accept new generation credit, debit and electronic cash cards. Although elimination of fraud and reduction of processing costs may drive the financial industry's transition to smart cards, these don't necessarily provide a convincing business case for retailers who must upgrade their payment terminals to accept the new technology. Indeed, many retailers feel, rightly or wrongly, that fraud is a banker's problem. Under those circumstances, it is clear now that even giving terminals away to retailers, at no cost, is not enough to persuade retailers to accept and promote an e-purse card.

On the other hand, the ability to run the retailer's loyalty program on the bank's smart payment card provides an added value function that directly interests the retailer. Loyalty technology has already proven to be a key catalyst facilitating the deployment of smart payment cards and terminals.

If you are planning to launch a new payment card product that competes against cash and cheques (whether off-line credit, debit or e-purse) and you want to make sure your smart card is used extensively by customers and is well promoted by retailers, it is crucial that retailers are able to run their electronic punch card loyalty programs on your card. This is simple common sense. Only then will they begin to accept the idea of paying to upgrade their terminals with chip card readers.

Multi-application smart card operating systems

Both Visa and MasterCard are promoting sophisticated card operating systems based on open, flexible and easy to program standards. Visa is pushing Java Card, developed by Sun, while MasterCard is behind Multos, developed by Mondex, a MasterCard subsidiary. Technically, the two platforms are very different, although their objectives are similar. Both platforms seek to establish a common standard so that application developers do not have to take into account details specific to each card manufacturer's proprietary operating system.

With Multos, all card manufacturers use the same operating system, analogous to the use of MS-DOS or Windows on all PCs. Programs are written in MEL, a language proprietary to Multos, compiled and loaded onto the chip. With Java Card, each card manufacturer can still offer their own, proprietary operating system. Programs are written in Java, compiled to byte code for efficiency, and loaded onto the chip. While a MEL program runs directly on top of the Multos card operating system, a Java Card program runs on top of a Java Virtual Machine.

The dream of some multi-application protagonists is eventually to see blank cards sold directly to customers who can then go to a kiosk or an automatic teller machine to select the applications they want on their card. They might choose a credit application from Citibank, a loyalty application for Starbucks, a frequent flier program with Delta, and perhaps a network access application or even a car personalization application that opens the car door and remembers how that customer likes the seats and rear-view mirrors positioned.

The technology will permit this. However, business issues have not necessarily been taken into account. For example, branding issues are rarely addressed. How would brands be managed in that type of environment? Will they simply disappear, as some people have suggested, or be replaced by a single, all-purpose, all-encompassing master brand, like 'Microsoft Card'? Will the card come with a bunch of peel-off decals for all the possible applications, leaving the customer to self-personalize the card? Or will stickers be provided by the kiosk once an application has been selected?

Who actually owns the card? Branding is closely linked to this issue. The prevalent thought seems to be that each application will be owned by a separate application provider, as in the PC environment. Your credit function doesn't work right? Call Citibank. You can't get your Starbucks electronic punch card to load on the card because the kiosk tells you the card is full? Call Starbucks.

What about privacy issues? Mixing commerce related items on a card along with non-commerce items will not be easy to get used to and will most likely trigger deep concerns about privacy. The idea today of using your Visa credit card to enter your office building is not self-evident. If you are pulled over for speeding and the police officer asks you for your driving licence, will you soon be presenting your Starbucks card? This is intuitively difficult to understand. Again, branding is at the heart of the problem. Was that really a Starbucks card you handed the officer? Or was it in fact a standard issue Microsoft smart card that you happened to personalize with a Starbucks loyalty program and sticker?

Suppose there was a paper-thin display on the front of every card? This would allow the user to see at all times what is stored in the chip. Variants of this technology are already available, but they require specialized read–write terminals that cost between $1000 and $2000 a unit. Current displays cannot be connected to the chip in the smart card, they can only be read and written to using a specialized device. However, we are not far from the day when we'll see inexpensive displays that are linked directly to the card's microprocessor chip.

Although the blank card concept, completely personalized by the user, is difficult to take seriously in the payment card and commerce environment, it does make a great deal of sense for secure access to networks, buildings and PCs, as well as for things like programming your car's seat positioning. The concept could soon prove to be a large market for multi-application smart cards.

The trick, as always, is to keep it simple. Most people still have trouble with their VCRs. What they want is to push a button and make everything work without being bothered by products that are too confusing and too complex. Devices that want to be all things to all people don't work. Remember the Newton – the fax, bleeper, daily scheduler, address book, note keeper, e-mail and cellular phone? Bill Gates describes the wallet of the future as a device that will replace your keys, credit cards, pocket change, driving licence, passport, health card and also carry photos of your children. Universal do-it-all devices like these are not likely to succeed simply because they have rarely succeeded in the past.

This century has shown that the trend is almost always toward specialized products for clear and easy-to-understand uses, as opposed to general purpose do-it-all products. When electricity was first being made available, electric motors were a novelty. The first home products offered an electric motor with lots of attachments. The market finally exploded once suppliers figured out it was better to put a small motor in each individual device. The same has happened with microprocessors. Each home today has perhaps a hundred microprocessors in thermostats, refrigerators, washing machines, VCRs, stereos and countless other objects.

Multos and Java Card both require a great deal of chip memory to manage overhead. Both card platforms are five to six times more expensive than traditional smart cards. Generally, the increased cost is said to be justified by the fact that each card will in fact replace several existing cards.

Existing smart cards cannot easily be reprogrammed after they have been manufactured. An independent application developer cannot simply create programs that are dynamically placed on the card at a kiosk or a retailer's payment terminal and executed at run time. This was of course a main reason calling for the creation of multi-application operating systems. But is this really a problem? How many card applications truly require their own executable code installed in the chip? Many of the most complex smart card applications use standard ISO commands that most card operating systems already support.

Loyalty software products that perform sophisticated customer behaviour analysis in real time are among the most complex applications available for smart cards. And yet, these applications only require the ability to securely read and write several data files in the chip. All existing microprocessor cards allow some form of read and write commands based on standards like ISO and EMV. So lack of a viable multi-application operating system has thus far not been an impediment to moving forward. Many useful services can be provided with off-the-shelf smart cards with proprietary operating systems that are already capable of supporting multiple applications concurrently, like credit, debit, e-purse and loyalty. And at a very attractive cost. Low cost chip cards capable of supporting a credit or debit function along with electronic punch cards are already available. They are ideal for applications that don't require a field programmable operating system like Java or Multos. So they cost just under $2, as opposed to over $10.

Cards with re-programmable operating systems, or multi-application operating systems, may one day also be available for $1 or $2, the price of some of today's standard off-the-shelf smart cards. When that happens, the opportunity to further develop intelligence at the chip level will become irresistible. Many additional uses of smart cards will become possible.

Smart cards will continue to evolve as computers and microprocessors have evolved. In the past, as whole industries moved to information technology, early adopters tended to fare better than their competitors who waited for the next generation of computer technology. The financial card industry does not need to wait several years for multi-application cards to drop in price. Much valuable work can still be done with current inexpensive technology. And the potentially large payoff of becoming an entrenched market leader through early adoption of game-changing technology provides visionary companies with a very strong incentive to move forward now.

Terminal standards have not kept up with card standards

As the Transaction Richness Quotient continues to increase, smart card and terminal hardware will require new capabilities. The industry has applied lots of brain power to solving interoperability and standards issues, primarily at the card level.

In general, all microprocessor cards today offer a standardized interface based on ISO specifications. Many cards also offer EMV compliance, which means that the things they do are done according to EMV definitions. Standard, off-the-shelf smart cards can often be integrated into a loyalty software system in several days, as opposed to the six to nine weeks required just a couple of years ago. By adapting to ISO and EMV standards on all card products, software applications like loyalty can be made to run concurrently with any payment application. Some readers may wonder how this can be done on cards that are not equipped with a multi-application operating system like Multos or Java Card. This is because much work has gone into developing software applications that simply rely on normalized file read and file write commands, avoiding fancy proprietary card level executable code. A big benefit of this approach is that the loyalty application can be used on all available cards, from the smallest card with 500 bytes to larger, more expensive cards offering 16 000 bytes of data space.

Terminals are another matter. Porting to a new terminal platform takes up to three months. Although this is still far better than re-writing such a complex application as loyalty (which has been evaluated at up to eighteen months by leading terminal manufacturers), our goal is to bring this time down to several weeks.

Smart cards are transforming the terminal landscape

In the past, card-processing systems concentrated intelligence centrally, on large mainframe computers that could respond instantly to hundreds of thousands of payment authorization requests. The retailer's terminal was a simple, inexpensive device that had virtually no intelligence. It didn't need to be intelligent. All it had to do was faithfully read the magnetic strip on the back of the card, understand the amount keyed in, dial up to its host, perform handshaking, transmit the purchase amount along with whatever was stored in the customer's magnetic strip, and wait for an authorization or a refusal. That's all. The terminal application was very simple to write and maintain. Each terminal manufacturer rewrote the payment application for each model offered. Because manufacturers typically offer half a dozen distinct models of terminals, each with a different operating environment, the application had to be rewritten for each of those. This might seem a waste, but the application was so simple that it was easier to rewrite it than to adopt a portable application software architecture.

Then smart cards came along, triggering a move to decentralized processing. Suddenly the terminal had to go from relatively stupid to relatively clever. Today, we are in the middle of a 'phase state transition', the term physicists use to describe the transition from gas to liquid or liquid to solid.

Some of the world's leading manufacturers have had trouble moving to the new mode of thinking based on significantly increased intelligence in the card and in the terminal. Many manufacturers still offer terminals that are quite basic. Some have even openly resisted the move to smart cards.

Phase state changes in the marketplace – or strategic inflection points, as Andy Grove calls them – are always a very dangerous moment for the current market leaders. The point-of-sale terminal market is not immune to this phenomenon.

Since the underlying philosophy of payment terminals is going through profound change, manufacturers cannot simply take their existing machines and throw in intelligence, which is unfortunately exactly what many companies have tried to do. Attempting to layer intelligence onto a product that cannot adequately support it, leaves the field wide open for newcomers with new products, redesigned inside out for new market requirements.

During the current phase state change in the terminal market, as terminals go from no intelligence to lots of intelligence, the current market leaders must depend and capitalize far more on their salespeople's personal relationships with customers, and far less on the company's technical products. At least until their engineers catch up and create new products that respond adequately to the market's requirements. This is a very vulnerable transition period for many of today's payment terminal industry market leaders.

French terminal companies have begun to capitalize on the vulnerability of companies like Verifone and Hypercom, respectively the numbers one and two US suppliers. Ingenico has created alliances with numerous terminal suppliers, including IVI/Checkmate, a US company that is the number three US supplier just behind Hypercom. Ingenico was one of the first companies in France to build smart card terminals. Its extensive experience – close to twenty years – has allowed it to build products that are now sold to other terminal suppliers that have not yet accumulated significant expertise in smart cards.

Large competitors like Verifone and Hypercom have been struggling to catch up. Verifone was bought by Hewlett-Packard and has turned more to developing e-commerce software. It remains to be seen if Hypercom will succeed in adapting to the new paradigm that relies heavily on intelligent terminals with sophisticated software applications. Recent announcements from Hypercom indicate a move in the right direction. For example, in March 1999 Hypercom announced a software development kit for third-party developers of point-of-sale applications on Hypercom's terminals. How long will it take their development kit to achieve the maturity of other products on the market?

When will palmtops and other very small PCs mutate into point-of-sale payment terminals? In the early 1980s the high end cash register market went through a similar mutation, where PCs became the heart of supermarket scanning terminals. At the time, manufacturers of hardware-based proprietary cash registers all complained in unison that PCs would never be sufficiently fast, robust or inexpensive. Cash registers required dedicated real-time operating systems that had to be programmed in hardware.

PCs are notorious for crashing at precisely the wrong time. Losing the price lookup server during a rush period is disastrous. This was solved by using dual server architectures. PCs are slow to boot up, typically taking up to a minute before they are ready to run, whereas point-of-sale scanning terminals have to boot up in several seconds. This was also solved, by putting the boot process in hardware. Numerous other seemingly unsolvable barriers were cleared in a few short years, resulting in PC-based scanning terminals that were as robust as hardware cash registers, less expensive, and an order of magnitude more powerful.

In less than ten years hardware cash registers had all but disappeared from supermarkets and department stores. They had been replaced by PC-based systems. The most successful suppliers concentrated on building open systems and encouraged the development of an independent software industry that could build innovative point-of-sale applications that could run on multiple hardware platforms.

Intelligence in the cash register meant companies like Wal-Mart could implement just-in-time inventory management right at the cash register, instantly taking into account scanning data as items are actually being purchased. More recently, Wal-Mart has even linked its scanning terminals directly into Procter & Gamble's ordering system, so products can be delivered to individual stores when they are needed. Fashion retailer Benetton claims that it no longer dyes its sweaters in bulk. It monitors detailed scanning data sent up by each point-of-sale scanning terminal every night, which allows it to immediately manufacture the exact colours and styles that customers are buying right now.

Increased intelligence at the supermarket checkout has been an important profit engine for many retailers for quite some time. It was an important source of differentiation for early adopters and has forced late adopters to join in and mutate their retail businesses in order to keep up. In 1988, Kmart had around 30 per cent of the discount retail market and Wal-Mart had about 5 or 6 per cent. Just a decade later, Wal-Mart had about 45 per cent while Kmart was down to under 12 per cent. This upset was completely unexpected. Wall Street analysts in the early 1970s were all predicting that Kmart was going to be the dominant retailer of the latter part of the twentieth century, because it was expected to crush Sears and be number one for a long time to come.[22] Point-of-sale technology allowed a relatively unknown retailer to race by and grab market share away from very powerful incumbents.

Sometimes, speaking with payment terminal suppliers, I can hear echoes of discussions I had over fifteen years ago in the point-of-sale scanning terminal industry. History repeats itself. The same revolution that swept through the point-of-sale scanning industry then will inevitably sweep through the payment terminal industry. It is already happening.

Terminal software design is a new business, independent of terminal hardware design and manufacturing

When terminals didn't need to know very much it was easy to write software for them. The payment associations and retailer acquirers provided specifications on their credit authorization protocols. These very basic protocols were programmed by a terminal manufacturer's engineers and were loaded into the terminal's hardware. Terminal software was simple to write and easy to maintain, since all the heavy duty work was being done at the host.

Today, terminal software is several orders of magnitude more complex than in the past. Payment software capable of performing an off-line authorization in no way resembles prior generations of terminal applications, nor does it resemble a pared down version of the host application. You cannot take an old credit card application and add lots of intelligence to it. Nor can you take a host application and somehow stuff it into a tiny payment terminal. You cannot even take some of the basic credit card authorization algorithms from the host and reuse them. Nor is the new generation payment application a hybrid between an old terminal application and an old host application. It is a completely new breed of software. The same is also true of an electronic purse application. And loyalty software is by far the most complex of all of these new generation payment terminal applications.

This fundamental change in the nature of terminal software has initiated a change in the value chain. The complexity of the application has given rise to specialist software companies that are independent of terminal hardware vendors. It is a consequence of the terminal industry's maturity. Just as hardware and software in the rest of the computer industry has been separated and companies must specialize in one or the other to succeed – and not both – terminal hardware companies and software application companies are distinct entities that must each specialize in their particular area. Dell chose to specialize in hardware, Microsoft chose software. Both companies show no signs of problems with their respective identities. Wall Street recognizes this and rewards it. IBM and Apple cannot decide whether they are hardware or software companies, so they do both.

Most terminal manufacturers today recognize this, which is why they now provide software development kits for third-party developers.

Unlike terminal manufacturers, software companies can develop their applications to be portable across terminal platforms that are not at all compatible. The terminal manufacturer has no incentive to do this, whereas the software vendor does. By making the application available on a larger number of platforms, the software vendor can sell the application to a larger number of customers. Customers can mix hardware platforms from various manufacturers and still use the same application.

Each terminal has its own proprietary operating system. Some manufacturers even offer a different operating system for each product line. More and more terminals are programmable in C, but that's still not always the case. Various flavours of C exist. Each terminal manages peripherals in a completely different way. Printers, displays, modems, card readers, internal clocks are all addressed differently. Memory paging is never done the same way. It took Welcome Real-time over 18 months just to define low-level drivers that were relatively universal. This was necessary in order to create a platform on which to build the loyalty application in a portable fashion. Moving to a new terminal essentially requires rewriting the low-level terminal drivers and then simply porting the higher-level application. Although this is new in the payment terminal environment, it has been standard software design practice in many industries for a long time.

Loyalty software capable of managing each retailer's electronic punch card program is very complex. It makes no sense to have terminal manufacturers rewrite the application based on detailed technical specifications. It is far easier to write the application once and port it to multiple platforms. Maintenance becomes much easier as well. Bugs are fixed in a master version, which is then simply recompiled for each platform.

Payment applications are also becoming more complex. Off-line credit and debit card payment applications are also being developed in the same portable fashion. Rather than have each terminal manufacturer write an application for each proprietary platform, third-party software companies are now writing portable applications.

Banks and merchants are increasingly demanding that terminals integrate numerous standards and norms that ensure long-term interoperability; card standards of course – like ISO, EMV, Java Card, Multos, etc. – but also general computing standards like the TCP/IP communications protocol used by the Internet, and standard languages like C and Java. Many of these standards are far more sophisticated than the simple protocols that payment terminals used to limit themselves to. Time and time again, in the mainframe industry, then again in the minicomputer and PC industries, the widespread adoption of mainstream standards has gone hand in hand with the creation of a strong, dynamic software industry separate and distinct from the hardware industry. In general, independent software companies are best positioned to develop and integrate mainstream software standards into various hardware platforms. This is now happening in the payment terminal industry.

The payment application and the loyalty application are becoming very closely intertwined. When an electronic punch card program runs automatically every time a card is used for payment, a single card insertion triggers two distinct applications that must both process the card and print receipts. The operation must be extremely smooth and seamless. It also must be performed in a handful of seconds, typically 3 to 5, counting from when the amount is entered to when the receipt is completely printed. For the sake of optimization, the payment and loyalty applications must be well coordinated. This is another reason why specialized software companies, rather than terminal manufacturers themselves, are best positioned to develop terminal applications. Getting a particular payment application to function smoothly and seamlessly with a particular loyalty application is lots of work. It is economically difficult for a loyalty software provider to justify redoing this work over and over again with multiple terminal manufacturers who have each designed their payment application in totally different ways. It makes much more economic sense to create a well integrated product once and simply port both applications together to multiple terminal platforms.

All new terminals will need to offer the ability to support partitioned applications so modifications to one application do not require recompiling all other applications and going through a complete recertification of everything in the terminal.

Terminals will also need to increase their processor speeds, which today are often limited to slow 8-bit microprocessors. Modems are another problem. A 1200 baud modem is usually fine for a short authorization request, which is why most terminals are still at that speed. Modems need to be much faster to better handle the large loyalty program parameter files sent by the server to the terminal at night.

Multitasking capabilities would be nice. The combined payment and loyalty process can be handled much faster if the whole terminal is not tied up while one process waits patiently for the payment host to return an authorization.

What about a standardized thermal printer with graphics capabilities? It would be great to have a receipt that showed a graphical representation of the customer's punch card, reprinted at each transaction, with Xs over each of the visits already counted. Some terminals are already capable of this, but since they have each to be programmed using non-standard commands, the resulting code is very difficult to port.

Still, much progress has been made. Java interpreters are being developed for terminals. Windows CE is also beginning to appear. Both of these developments will go a long way toward creating useful solutions once they are made sufficiently fast, robust and inexpensive.

And of course, even with all of these necessary improvements, terminals still must cost under a few hundred dollars.

Evolutions in smart card and terminal platforms: three fundamental concepts

1. Loyalty software products that perform sophisticated customer behaviour analysis in real time are among the most complex applications available for smart cards. And yet, these applications only require the ability to securely read and write several data files in the chip. All existing microprocessor cards allow some form of read and write commands based on standards like ISO and EMV. Many useful services can be provided with off-the-shelf low cost smart cards with proprietary operating systems that are already capable of supporting multiple applications concurrently, like credit, debit, e-purse and loyalty.

2. In the past, the retailer's terminal was a simple, inexpensive device

that had little or no intelligence. The terminal application was very simple to write and maintain. Then smart cards came along, triggering a move to decentralized processing. Suddenly the terminal had to go from relatively stupid to relatively clever. Today, terminal software is several orders of magnitude more complex than in the past.

3 Just as hardware and software in the rest of the computer industry has been separated and companies must specialize in one or the other to succeed – and not both – terminal hardware companies and software application companies are distinct entities that must each specialize in their particular area.

8

THE BUSINESS CASE FOR SMART CARDS

❖

Those cards are expensive and require specialized hardware in every store. Where is the business case? They'll never take off.

Banker speaking about magnetic strip cards in 1975

In 1985, Citicorp launched an ambitious project to put loyalty cards in the wallets of 40 million US consumers. By means of supermarket scanners, linked to a central management system, they hoped to build up a wealth of information regarding the buying habits of cardholders. Through this system, consumer goods manufacturers would be able to offer discount coupons to buyers of competing products, or discounts to customers who buy a certain quantity of a given product over a predefined period.

How did it work? At home, the consumer received promotional offers, such as 'Buy 3 boxes of Captain Crunch cereal and get a $1.50 discount'. The customer's card was swiped at the supermarket cash register, so the customer's identity could be recorded together with the items bought. After the information had been analysed, the customer received a $1.50 coupon in the post.

Five years and 200 million dollars later, the program was only able to attract 2 million cardholders, or 5 per cent of its target. The centralized approach turned out to be too heavy to manage. Information came too late, often by several months, and included massive data, too unwieldy to analyse. Customers received confirmation of acquired benefits long after they had made a purchase. Showing the card at the supermarket cash register had no direct immediate link with obtaining the discount. The project demonstrates that centralized systems that rely primarily on direct mail are too slow and cumbersome, not to mention astronomically expensive, for heavily customized marketing programs.

Real-time information delivery costs less

A smart card solution often turns out to be less expensive than a magnetic strip card solution, when one globally considers infrastructure costs as well as operational costs. This is due to the inefficiencies of the magnetic strip card program: its entirely centralized system relies heavily on post to communicate between the cardholder and the issuer.

With a smart card, advantages are loaded directly into the card's memory chip. Information is readily available to the cardholder through a statement printed at the payment terminal, providing a snapshot of the chip's contents. The statement provides information regarding the chip's prior loyalty balance, the benefits obtained and the new balance.

The customer is immediately aware of the card's status and contents. In this way, direct marketing budgets are not wasted sending monthly statements to already loyal customers; they are focused towards regaining customers who are no longer loyal. The relationship with the customer increases in quality as it becomes more personalized.

In general, the costs to set up and manage a traditional magnetic strip-based loyalty program are so high that little money remains to finance customer benefits. The major costs of a smart card program are the cost of the card itself and the actual customer benefits, unlike a magnetic strip card program where the majority of costs are dedicated to program management and administration.

Smart cards eliminate the need to post the twelve statements a year that are normally required. Since the additional cost of a smart card, as compared to a magnetic strip card, is approximately equivalent to the cost of 3 mailings, the overall smart card solution turns out to be far less expensive. The smart card market follows general computer technology trends where each new generation costs less. New cards will soon be available at around US$1 – that way, the price difference comes close to the cost of a single mailing.

Because some aspects of customer behaviour are recorded on the card's chip each time it is used, the system is able to instantly recognize and reward frequent users and big buyers, by delivering rewards based on recency, frequency and monetary value behaviour patterns.

Using a simple bank payment terminal, the retailer is able to reward precise customer behaviour, for example by offering one bonus point at the customer's first visit, three points at the second visit, and five points at the customer's third visit each month.

Smart cards also significantly reduce telecommunications and centralized data-processing costs. The chip's built-in security and intelligence allow for doing away with real-time connections at the point of purchase. Transactions are uploaded into the central system at night, when communication costs are low and data-processing equipment is unused. Very cheap infrastructures, such as the Internet, are becoming available for this type of staggered transmission.

AOM French Airlines' smart card program is targeted toward frequent fliers, essentially replacing the company's existing 'Carte Capital' and extending it to new passengers. Since the program was launched, AOM has registered a high demand for new membership, well beyond forecasts.

AOM's management was surprised at the cost reductions. 'Thanks to instant statement printing at each transaction, our smart card system allows for improved service to our customers, while at the same time doing away with frequent statement mailings and the related call center necessary to answer the inevitable questions concerning these statements,' said AOM's General Manager, Jean-Marc Janaillac. 'For us, the smart card clearly represents major cost benefits when compared to the magnetic stripe card.'

AOM's ticket and check-in counters were equipped with 85 bank card-type payment terminals. When the passenger checks in, the 'Carte Capital' smart card is credited with a number of points based on several parameters such as destination and class. Promotional campaigns can be launched at any time to offer bonus points – for example, during the month of May 1997, AOM doubled the number of points credited on all domestic flights. The total number of points is stored in the card's microprocessor chip. At any time and at any AOM sales counter, acquired points can be debited in exchange for free tickets or instant upgrades.

AOM's customers benefit because they always know how many points they have. Points are credited and debited instantly each time the card is presented at the terminal. Points stored in the card's electronic purse can be exchanged for rewards at any time and at any terminal, with AOM or with other participating service providers. The terminal prints a detailed loyalty statement after each transaction, showing the complete status of the cardholder's 'Carte Capital' account (prior balance, points earned with this transaction, new balance, promotional and incentive messages, etc.).

AOM uses the loyalty statements as a direct marketing tool. Printed receipts eliminate the need to post costly account balance statements, resulting in significant savings on direct mailing expenses. The program functions on a very light and cost effective IT infrastructure.

With smart cards, systems can be designed to be scaleable, robust and inexpensive to manage. By adding intelligence at the card and terminal level, and not concentrating it only at the central system, a single program architecture can be used for small programs just as well as for large programs with tens of millions of cardholders and hundreds of thousands of merchants.

The use of smart cards means that all transactions are processed completely off-line with no requirement for on-line authorizations or updates. Each retailer's own electronic punch card program can run independently of a central server. This means that in reality, servers are optional and can in some cases be completely avoided.

When a server is used, nightly data collection procedures can rely on standard, mainstream communications protocols like TCP/IP and e-mail, which means that common Internet access providers can be used for data exchange. This eliminates the need for the program operator to manage rooms full of modems which are instead owned and run by Internet access providers. Terminals call up a local phone number at night that rings the access provider. The terminal presents its log-in and password, then proceeds to package its transaction file as the text of a standard e-mail message. The message is addressed to the server's e-mail box and is sent off through standard e-mail protocols. Terminals generally need to transmit transaction files of several hundred thousand bytes, a fraction of what Internet access providers normally expect with other Internet usage, like Web surfing. Furthermore, terminals send their data nightly, when most Internet access providers are least busy. Internet access providers generally have lots of unused bandwidth at night, dimensioned for a far greater exchange of data. Sending data back and forth between the server and the terminal via e-mail means that the server does not need to be connected at night. The server simply logs on in the morning and consults its incoming e-mail containing each terminal's data collection file.

This type of highly efficient and inexpensive data exchange is made possible because of the smart card's off-line capabilities. Existing credit card systems are based on frequent short exchanges of data between the terminal and a central payment authorization system. Typically, the terminal requests a credit authorization from the host and waits for an immediate response, generally something that boils down to yes or no. Tiny chunks of data must be exchanged extremely fast. Speed is far more important than flexibility. Credit card processing protocols are designed to be fast and to virtually never change in format or content. In many cases, the low levels of the protocol, those closest to the hardware, are developed with a knowledge of the actual nature and format of the messages exchanged between the terminal and the host. This makes the connection happen very fast, but it also makes it very difficult to modify or enrich the information carried by each message. If new pieces of data are to be included in the transfer, the lower levels of the protocol must be modified. This is very complicated and requires specialized programmers. With smart cards, everything is reversed. There is no need for instant turnaround on short messages sent during the day. However, there is a need to exchange much larger files at night, both the transaction data going from the terminal to the host as well as detailed loyalty parameters and marketing program updates going from the host to the terminal. Because new functionality is constantly being developed in the card and terminal, the format and content of data collection files are under constant evolution. The data exchange protocol must be highly flexible. Any modifications to the nature of the data must happen without impacting the low-level protocol.

With standard communication protocols like TCP/IP (the Internet's protocol) and e-mail, program operators can easily administer international programs, since terminals anywhere in the world can connect up to a local Internet access provider and address their e-mail message to a server on the other side of the planet. This can give birth to new, worldwide loyalty program service companies. There is no reason why a server must be located in the same geographic region as the terminals it administers.

Why not let a terminal send its data collection file to several servers at the same time? This can be done very easily by simply identifying several server e-mail addresses as the message's destination. It is an elegant way to easily allow multiple service providers to concurrently monitor a terminal's activities. One provider might be the program operator, another might be the store's central marketing department, while a terminal maintenance organization might use automatically generated remote diagnostic information to follow the terminal's day-to-day operations and intervene before significant problems occur.

Servers can be configured to receive full detail concerning each transaction, or simply a summary of each terminal's activity. When a smart card loyalty program is used in conjunction with payment cards the payment transaction data is already available via the payment acquirer. So the option to upload only a summary of loyalty activity eliminates duplicate data transfers and greatly increases the number of retailers that can be managed with a single off-the-shelf server.

A retailer can choose to define his own loyalty program directly on a PC with easy to use PC software. Parameters are then loaded to the terminal via a smart 'Merchant card'. The retailer can define programs offering a free item or a discount after a specific number of visits or dollars spent in a given period at the retailer's store. Loyalty program reporting can be made available directly at the point-of-sale terminal.

If the retailer doesn't have a PC, or simply doesn't want to define the program's parameters himself, he can choose to let his terminal installation and maintenance organization manage the parameters. The terminal maintenance organization's representative would then use the same PC software on a laptop to define parameters, directly at the retailer's site, and use the retailer's smart card to load parameters to the terminal.

Often chains will want to run chainwide loyalty programs, rewarding customers based on purchases cumulated across all stores belonging to the chain. Individual stores can continue running their own local electronic punch card programs, in addition to programs defined centrally by the chain. Chain wide punch card programs offer customers a free item or a discount after a specific number of visits or amount of money spent in a given period across all stores participating in the program. Each chain's server can receive detailed program data from each terminal, so reports can be generated by the chain's central marketing personnel.

What types of loyalty programs are best managed using smart cards?

In virtually all market sectors, the trend is going away from mass marketing and simple programs limited to a single store, toward one-to-one marketing and programs with multiple stores and chains. This general trend favours smart card technology. Although all retail categories will eventually be affected by this trend, some will be required to move earlier than others. Smart cards pick up where centralized systems cannot perform adequately. The two-by-two matrix in Figure 8.1 shows the types of programs that are best addressed by smart card technology today.

Petrol stations are a typical example of a retail category where smart payment cards provide a clear benefit. Customers stop at the closest station when the tank is empty. The number of petrol stations makes it unwieldy to have an up-to-date customer database available instantly at the point of sale. Fast food chains and coffee shops are other good examples. Customers typically eat at many different restaurants belonging to the chain. Transactions must be fast, so connection to a central database is not viable. All independent retailers that already offer punch cards to their frequent shoppers are also likely candidates for smart card-based loyalty programs running on new generation credit, debit and e-purse cards issued by financial institutions.

What about retail categories where smart cards do not necessarily provide a clear benefit? Supermarkets are a prime example. Customers tend to shop at the same store where sophisticated point-of-sale systems are capable of storing a customer database on the point-of-sale local area network, so customer information can be made available instantly. Smart

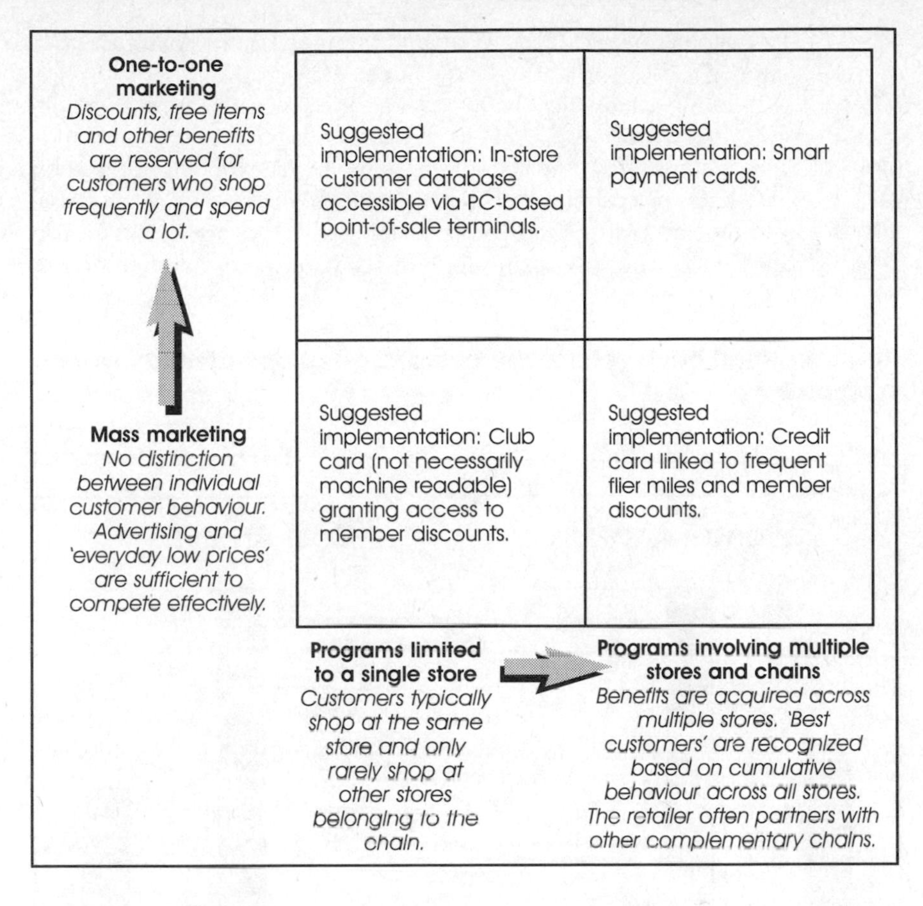

Figure 8.1 Loyalty programs best suited to smart cards

cards will be used at supermarkets when customers begin receiving electronic coupons from the Web, storing them in their credit cards, and when consumer goods manufacturers begin offering their own electronic punch cards running on smart Visa credit cards.

Video rental stores are another example. Checking out a video requires a detailed customer account. The same computer that manages the account can easily offer things like an eleventh video free. However, if a particular chain's customers tend to pick up and drop off videos at numerous different locations, smart cards can help simplify accounting procedures.

Here are a few key questions to ask retailers in order to evaluate the need for smart cards:

- Do you already offer loyalty programs such as 'spend *X* pounds during the month, get a free item'?

- ❍ Do customers typically shop at multiple outlets belonging to your chain?
- ❍ Is building chainwide loyalty as important as loyalty to a single store?
- ❍ Even if single-store loyalty is sufficient (i.e. store does not belong to a chain, or customers do not typically shop at multiple outlets), is instant access to a customer database at the point of sale too difficult or expensive to implement for your store environment?

How can smart cards reduce the overall cost of operating a loyalty program?

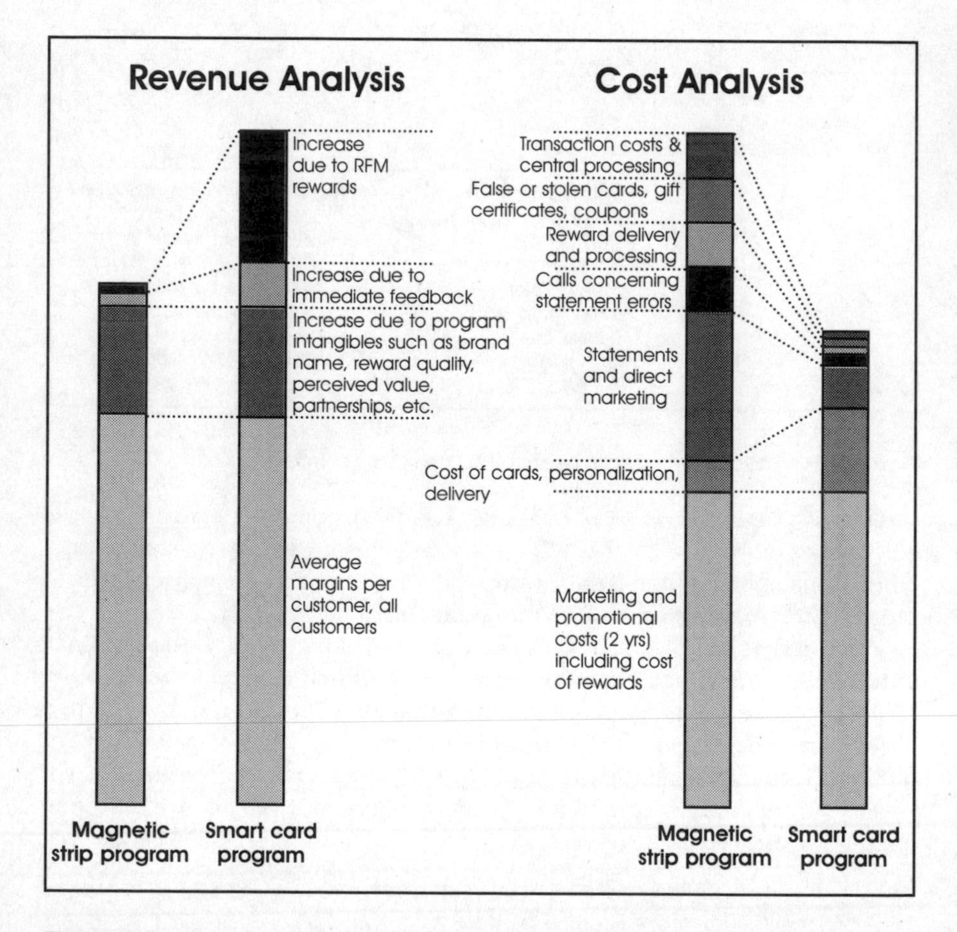

Figure 8.2 Smart card business case block diagram

With a smart card, loyalty statements are printed at the payment terminal, containing the card's prior loyalty balance, the benefits obtained and the new balance. Marketing budgets are not wasted by sending monthly statements to already loyal customers; they are focused towards regaining customers who are no longer loyal.

Benefits are easy to acquire and redeem. A customer who earns 45 points in the morning by buying a new tennis racket, for example, is able to use points for a free video ticket that same evening. Reward processing overheads are eliminated. (See Figure 8.2 for a comparison of costs with magnetic strip cards.)

In general, a smart card costs less than:

- three statement mailings
- one call concerning a statement error (such as last week's Paris–London round trip not showing up in the statement)
- fees for processing and handling a single reward (such as a voucher, bonus cheque or gift certificate, or processing of a selected catalogue item).

How can smart cards increase revenues generated by the loyalty program?

Revenues are increased by rewarding frequency and spend. (See Figure 8.2 for a comparison of revenues with magnetic strip cards.) The success of the program depends on several criteria:

- *Reward increased spending* of 30 to 35 per cent above average (i.e. the fourth visit of the month when the average is three, or total spend of $200 when the average is $150).
- *Reinforce positive behaviour* through immediate feedback (when a reward threshold is met, the customer is immediately informed at the point of purchase).
- *Eliminate barriers* to program success by simplifying administrative procedures such as the need to post letters, collect coupons or fill out forms.

A conservative assumption is that only 30 per cent of cardholders increase spending in order to obtain the reward: 10 per cent of cardholders spend *15 per cent more* than the required amount; 10 per cent of cardholders spend exactly the required amount; 10 per cent of cardholders spend *15 per cent* less than the required amount.

In this simplified model, total revenue over all cardholders is increased by well over 10 per cent.

Table 8.1 Comparison of large loyalty program costs

	OPTION 1 Program managed centrally through direct marketing	*OPTION 2 Program managed centrally through on-line transactions*	*OPTION 3 Program managed through store's local database*	*OPTION 4 Program managed off-line through smart cards*
Technology	Magnetic strip	Magnetic strip	Magnetic strip or barcode	Smart card
Physical location of the customer's account information	Central database	Central on-line database	Store POS server + central database	Smart card + central database
Connection to the processing centre	Nightly	At each transaction	Nightly	Nightly
Loyalty account statement available at store cash register	No	Yes	Yes	Yes
Time between customer purchase and statement issue	Several weeks	>1 minute	Immediately	Immediately
Call centre required to respond to questions concerning customer statements (usually misunderstandings due to the time lag between actual purchase and statement issue)	Yes	No	No	No
Ability to use card across the retail chain	Yes	Yes	No	Yes
Ability to use card across multiple retail chains	Yes	Yes	No	Yes
Ability to redeem points for rewards by simply presenting the card at a store cash register	No	Yes	Yes	Yes
Ability to redeem points for rewards outside the retail outlet, through participating retaillers, by simply presenting the card at a cash register	No	Yes	No	Yes

	OPTION 1 Program managed centrally through direct marketing	*OPTION 2 Program managed centrally through on-line transactions*	*OPTION 3 Program managed through store's local database*	*OPTION 4 Program managed off-line through smart cards*
Ability to use card in vending machines and other self-service devices (parking meters, telephones, games ...)	No	No	No	Yes
How points are typically redeemed by the customer	Phone request, mail delivery	Upon card presentation	Upon card presentation	Upon card presentation
Type of retailer equipment required	Simple payment terminal	Simple payment terminal	PC-based POS scanning terminal	Simple payment terminal
Cost of retailer equipment	$500.00	$500.00	$4 000.00	$500.00
Card costs (per card, based on qty of 500,000 units)	$0.50	$0.50	$0.50	$2.00
Printing costs (guide book, rewards catalogue, etc.)	$0.60	$0.60	$0.60	$0.60
Annual database management costs (per cardholder)	$0.50	$0.50	$0.50	$0.50
Statement issue costs (per statement)	$1.00	$0.20–$0.40	0	0
Cost of redeeming points toward rewards	$8.00	$0.20–$0.40	0	0
Call centre costs, per call	$4.00	NA	NA	NA
Estimated total cardholder costs over 2 years	$30–$40	$20–$30	$2	$4
Summary	High costs; heavy infrastructure; time lag between actual purchase and statement issue	High costs; heavy infrastructure; increased processing time at the cash register; risk of server failures	Heavy infrastructure (requires a high-end PC-based POS scanning terminal); lack of openness and flexibility	Relatively low costs; light infrastructure; maximum flexibility

Because purchases tend to move toward higher margin items, total margins are increased by 15 to 20 per cent.
Over two years of operation, traditional programs cost five to ten times more than smart card-based programs. Traditional programs cost less up-front, since magnetic strip cards cost very little, but are far more expensive to operate. Savings generated to the retailer over a period of two years total between $15.00 and $35.00 *per cardholder*.

Loyalty provides new revenue streams for smart payment card issuers

This cost discrepancy provides an opportunity for smart card issuers to generate new revenue streams by charging retailers for the ability to run their private electronic punch card loyalty programs on the issuer's smart payment card. By eliminating the need for the retailer to issue and manage private cards, significant savings can be achieved for the retailer. By invoicing only a small portion of these savings in the form of transaction fees, loyalty can in many cases pay for the smart payment card in full.

Retailer loyalty programs running on private cards generally cost the retailer up to 1 per cent of sales. Over half of the program's costs are related to infrastructure and operational expenses, less than half of program costs are the actual rewards given to customers. This means that smart credit card issuers can charge retailers up to 0.5 per cent of sales for each transaction linked to an electronic punch card program. This 'loyalty' fee is in addition to standard payment processing fees charged by the retailer's acquirer.

Loyalty brand managers can generate additional revenue streams by licensing their brand to other card issuers.

Loyalty can reduce the cost of acquiring new cardholders

For many card issuers, acquiring a new cardholder often costs up to $100, which is mainly marketing costs related to countless expensive mailings of brochures and application forms. Once a new cardholder is signed up, the battle is not over. The cards need to be used frequently. Finally, a large number of customers do not renew their cards upon expiration, meaning that card issuers must begin again from scratch.

Retailers that run electronic punch card programs will actively promote the loyalty brand they have adopted, for example by placing in-store signs offering a free sandwich the fourth time customers pay with their 'Instant Rewards' card. When a retailer's loyalty program depends heavily on cards issued by financial institutions, the retailer will probably do everything possible to help customers become cardholders, for example by offering cards directly at the store counter and systematically asking for the card. You can also be sure that retailers will go out of their way to make sure the cards are actively used.

All card issuers, regardless of whether they own a loyalty brand or license it from someone else, can benefit by earning a higher share of the retailer's transactions and by enjoying reduced cardholder acquisition costs. Smart card loyalty technology allows early adopters to win out over their competitors, while at the same time allowing the overall financial card market to gain ground against other payment methods like cash and cheques.

The business case for smart cards: three fundamental concepts

1. A smart card solution often turns out to be less expensive than a magnetic strip card solution, when one globally considers infrastructure costs as well as operational costs. Advantages and behaviour data are loaded directly into the card's memory chip, so information is available to the cardholder through a statement printed at the payment terminal. Costs are reduced by eliminating the need to post loyalty statements and process paper coupons and vouchers. Smart cards also significantly reduce telecommunications and centralized data-processing costs.
2. Smart cards are ideal for programs involving multiple stores and chains and those which need to selectively deliver discounts, free items and other benefits to customers who shop frequently and spend a lot. The smart card becomes especially attractive when 'best' customers are recognized, based on cumulative behaviour across multiple stores, and rewarded instantly.
3. In addition to boosting the use of their payment cards, banks generate new revenue streams by charging retailers for the ability to run their private electronic punch card loyalty programs on the issuer's smart payment card. Many retailers will pay up to 0.5 per cent of sales for each transaction linked to an electronic punch card program. Loyalty brand managers generate additional revenue streams by licensing their brand to other card issuers.

9

KEY FACTORS FOR A SUCCESSFUL SMART CARD PROGRAM

❖

Flowing water avoids the high ground and seeks the low ground ... avoid difficult methods and seek easy ones. Do simple things well and quickly.

Sun Tzu

Smart card pilots should certainly be used to test technical choices, but just as important, if not more so, they should also concentrate on validating the overall value proposition. Technology can be tested in laboratories far more effectively and quietly than in live pilots that mobilize hundreds of retailers and tens of thousands of customers. The value proposition is much harder to test and even harder to fix.

Smart cards run completely off-line. Intelligence at the card and terminal level means the central system is much easier to manage. Server and telecommunications bottlenecks simply do not exist, so a laboratory test with a few terminals and a dozen cards will often produce essentially the same technical results as a test with hundreds of retailers and thousands of cards. Stress testing a smart card system can be compared to stress testing Microsoft Word running on PCs. How many people can run Word on their PCs at the same time without running into a bottleneck? A hundred people? A thousand? A million? Twenty million?

Sounds like a silly question? Some smart card pilots are performing tests that are essentially equivalent.

Laboratories are for testing the technology, pilots should concentrate on testing the value proposition.

The Manhattan e-purse project generated a wealth of information on what retailers and customers would or would not accept in a new payment card product. The test did provide some technical information, but not much more than what could have been learned in a laboratory environment. What is far more important is the greater level of understanding that the industry now has concerning e-purse implementations. The test showed that operating procedures for retailers and customers must be very simple. Customers have trouble accepting the idea of downloading cash to their cards from an automatic teller machine rather than simply obtaining a few banknotes. Banknotes can be used everywhere, the e-purse cannot. Retailers had trouble understanding the need to learn separate payment processing procedures for Mondex and Visa Cash, which function very differently in the way they are accounted for and how data collection is performed. This e-purse test in particular was extremely useful for the market. Since the test was completed, a large majority of banks that had previously been considering a move to e-purse technology shelved those projects after concluding that the market was not quite mature enough yet. This cleared the way for more pragmatic projects based on simple, easy to understand uses of smart cards to improve the banking industry's core credit processing business.

Create a clear and simple value proposition for retailers

If you are considering launching a smart card pilot, common sense suggests that you make sure retailers and customers both see very significant value in your card program.

What are the advantages for retailers? How will they be persuaded to upgrade their existing payment terminals with smart card readers?

Are you sure they will promote the use of your card in their store? If not, what can you do to make that happen? That's not all. You have to be 100 per cent sure that their needs are so well addressed that retailers will want to play a proactive role in your pilot, pushing your card over all other available payment methods.

Retailers want things to run real smoothly, without having to do much in the way of pushing buttons and analysing reams of data. Surprisingly enough, most retailers are not attracted to the idea of knowing who their customers are and what they buy. Only very large, centralized retailers have made moves toward data mining systems. Even then, attempts at data mining have been very timid. Retailers are pragmatic. Does the system make perfect sense? Will it help deliver special promotions to my best customers, so that I can concentrate valuable rewards on customers that really do spend a lot in my store? Does it run on its own without my having to become a database expert (and without my having to hire one)?

Suppose I own a fast food restaurant franchise that is part of a big chain. I would like to give a free value meal to customers that have already bought four other value meals this month at my restaurant (I don't really care what customers are buying at other restaurants belonging to the chain). If I know that only my customers get the free value meal, it is very easy for me to calculate a return on investment. I don't need to know each customer's name, address and purchase history. I just need to know how many free value meals I gave out and whether or not my restaurant's overall sales have increased.

Make it simple for customers to participate

What are the advantages for the customer? Customer reward programs often get bogged down by administrative details, and unfortunately this happens quite easily. Many marketing people can invent a neat, exciting rewards concepts that builds on a brand's value. It is much harder to carry the concept through to implementation without it becoming transformed into something far less exciting, thanks to Byzantine procedures that even Kafka would have been impressed with.

Keep one eye on the high-level objectives, the other eye on the details. Dive in to the practical 'how does it work' aspects, but never take one eye off your core objective. Find ways to eliminate all the complicated and messy details that are visible to customers. If you cannot eliminate certain details, hide them. Make them your problem, not your customers. This will make your offer significantly more complex overall, but you will gain a wealth of appreciation from customers by providing them hassle-free services.

Customers sometimes feel that procedures are complex because companies want to make it harder to acquire the reward. Here's a complaint many marketing professionals have heard: 'If everybody got the reward, the company would go out of business'. Or another: 'They're just misleading customers. They get people to buy something for the gift advertised on the box, then they make it impossible once you read the fine print.'

I don't believe most companies do this on purpose. Those few that do need to be aware that customers are not naive.

Provide numerous opportunities for retailers and customers to meet and use your card

Cardholders with chip cards must represent a very significant proportion of a retailer's customer base. A loyalty program that only concerns a minute percentage of customers will have a negligible impact on the retailer's total revenue.

Likewise, cardholders must be able to obtain electronic punch card rewards at a significant number of high-quality retail outlets. If too few retailers offer rewards (or if participating retailers are the least attractive in their respective categories) the card has little perceived value beyond that offered by a standard payment card.

Create a loyalty brand and prepare to license it to other card issuers. If a punch card loyalty brand like 'Instant Rewards' already exists in your market, join it. It is one of the most effective ways to maximize retailer and cardholder participation.

Retailers should offer significant rewards ... and actively promote them

Rewards offered by retailers to their loyal customers should be significant. A reward representing a deferred discount of 1 or 2 per cent after spending a hundred dollars or so is a poor way to thank loyal customers. Loyalty programs that offer all customers a fixed number of points for every dollar spent are forced to set the percentage low. Punch card-type loyalty programs focus on rewarding customers only after they have effectively spent a certain amount in a given period. A far greater reward value can therefore be provided to loyal customers, since rewards are not wasted on other customers.

Advertising, marketing, mailings, in-store signing etc., should be significant enough to allow customers to understand the benefit of using their new card.

Cardholders should be able to easily and quickly know where they can use their card, so they can choose participating retailers over non-participating retailers without going through an obstacle course.

Maximize the customer's motivation to make the next purchase

Retailers should increase the value of rewards as a customer spends more or shops more frequently. Airlines are successful with a variation of this by offering 1 mile for every pound of economy class travel, 1.5 miles for business class and 2 miles for first class. Make the next purchase increasingly more valuable to the customer. This locks the customer in and increases switching costs over time. It also reduces rewards paid to casual buyers, which then permits even greater rewards to frequent buyers without increasing the overall marketing budget.

Oil company Total developed a loyalty card in the early 1990s that offers an added advantage in addition to the usual 1 point for every 10 litres of fuel purchased. If your car breaks down, for any reason, within 15 days after filling up at Total, on-the-spot assistance will be paid for by Total. The advertising campaign was clear and simple. The message was well received, especially among Total's female customers, a primary target for the company's loyalty card program. Fifteen days is a normal period of time between two fill-ups, so the motivation to go back to Total is strong. Why let an insurance policy expire when all you have to do is go back to Total for fuel?

Total's program can be improved. Today, because the card does not have a chip, customers must pay for the towing services out of their pocket, keep all receipts, send them in to Total with their card number and wait for Total to complete the process and send back a cheque. It would be easier if the garage mechanic simply placed the customer's card in a reader, verified that assistance was paid for by Total, and be done with it.

Retailers should offer rewards that enhance the value of their product or service

The GM card enhances the value of the car, because it allows customers to use their accumulated loyalty points to move up to a higher product category without paying more. An airline's frequent flier program enhances the value of the airline because passengers can upgrade for free, or travel with their spouse for the price of a single ticket.

Retailers that give punch cards already follow this guideline intuitively. A sandwich shop will typically offer a free sandwich after buying several. In some cases, offering a complementary product like a free drink or desert could be an even better way to enhance the value of the core product. Plus, the perceived value of the retailer's products and services is greater than the actual cost to the retailer, therefore the loyalty program is more effective and costs less.

Programs offering rewards for other companies' unrelated products or services might have a promotional impact on customers, but they have questionable loyalty impact. Some programs might even damage the retailer's brand. Several years ago a credit card company in France launched a loyalty program offering points that could be collected and exchanged for fuel vouchers. Hard to believe but true. The program bored itself to death. Most retailers were smart enough not to get their brands mixed up with such an unexciting concept.

Avoid mileage programs – or do them right

If you are able to add significant value to points issued by retailers, it might make sense to launch a mileage program. If you simply act as a loyalty transaction acquirer – invoicing retailers for points they give customers and reimbursing them for points they accept as payment – then you are probably taking value out of the system, in which case the mileage program will fail.

In most cases, you are far better off allowing retailers to use your card infrastructure to run their own private electronic punch card programs.

Build on habits that customers are already comfortable with

Are you asking customers to radically change their current behaviour? It's taken them a long time to develop a smooth relationship with credit cards, automatic teller machines and payment terminals. Requiring them to learn new concepts like how to reload their electronic purse at a kiosk or at an automatic teller machine raises very powerful roadblocks that you will have to overcome. Many of these are needless. With some common sense and forethought virtually all such barriers can be eliminated.

A popular piece of folk wisdom says 'never simultaneously switch jobs, towns and spouses'. Change one item at a time. Adjust yourself then go on to the next.

Customers are familiar with using their cards to pay at point-of-sale terminals and to withdraw cash at automatic teller machines. Retailers are used to punching a purchase amount on their terminal's keyboard and swiping the customer's card. That's the starting point. Early implementations should stick to these basic habits as much as possible. Systems must be simple to use and must not require significant changes in how retailers and customers currently perform payment card transactions. Additional keystrokes at the terminal must be avoided, as well as operations like reloading an electronic purse, or performing maintenance on the card's contents at a kiosk.

Once retailers grow accustomed to inserting a smart card, and customers are used to the idea of having information dynamically updated and stored in the chip, the market will be ready to learn new habits, for example using a kiosk or a PC to see what information is stored on the card. Then, once people are familiar with that, you might eventually be able to persuade a few early adopters to perform a completely new type of behaviour: adding coupons to their card or reloading an e-purse.

Build on what people are already familiar and comfortable with. Don't rush them into whole new ways of doing things.

Table 9.1 shows how new card-handling methods might be added in an evolutionary manner that builds on successive waves of customer and retailer adoption.

Table 9.1 New card-handling methods

	Today	*1st generation evolution*	*2nd generation evolution*	*3rd generation evolution*
Payment operation, customer's point of view	Present card, sign receipt	No change	Enter PIN	No change
Payment operation, retailer's point of view	Swipe card, enter amount	Insert card	Request PIN	No change
Customer ATM operation	Insert card, enter PIN, choose amount	No change	View chip contents	Add coupons to chip, reload e-purse
Home PC operation	Enter card number and expiry date	Insert card in reader, enter PIN	View chip contents	Add coupons to chip, reload e-purse
Use of credit card chip to store *non-commerce* type data (i.e. network or building access, prescriptions, etc.)	Not supported	Wait until customers have adjusted to the idea of smart cards storing *commerce*-related data like payment, discounts and coupons		

Loyalty software features to look for

The following list shows the typical features to look for in a smart card loyalty system. Use it as a checklist to compare multiple software products.

General loyalty functions

- ❍ Electronic punch cards based on customer's frequency and cumulative purchase amounts
- ❍ Electronic punch card expiration
- ❍ Loyalty point issuing based on customer's frequency and cumulative purchase amounts

- Loyalty point redemption as a payment method
- Loyalty point expiration
- Ability to invoice/credit retailers for points
- Loyalty statement printing, providing information regarding the chip's prior loyalty balance, new benefits obtained and the current balance
- Ability to concurrently run multiple e-punch card programs in the same terminal

Customization/parameterization

- Definition of retailer terminal parameters for a loyalty program dedicated to that retailer
- Ability of a retailer to mix his own program in parallel with the chain's program
- Ability of a program to be customized for a single store

Security

- Loyalty can be locked without affecting other applications
- Maximum limit of manual points issued by retailer/supervisor
- Security of retailer specific data from unauthorized access, using Triple DES or RSA
- Display points issued/redeemed after each transaction
- Ability to prevent/detect fraudulent points
- Ability to set maximum number of points which can be uploaded by terminal
- Terminals must upload information before overwriting transaction logs

Data transfer

- Transmission of transactions to the card program management system, via Internet
- Download of parameters to terminal via a smart card
- Reception of parameters configured centrally on the program management system, via Internet

Reporting

- Reporting of points issued/redeemed by retailer
- Reporting of punch card activity by retailer
- Report data available directly at the retailer's terminal through flash reports

Product maturity

- ❍ Date of first implementation
- ❍ Total number of cards issued
- ❍ Total number of terminals installed
- ❍ Current version
- ❍ Number of employees fully dedicated to loyalty application

Scaleability

- ❍ Are loyalty transactions completely off-line?
- ❍ How many retailer programs can co-exist in 500 bytes of card memory?
- ❍ How many distinct retailer programs can exist system-wide?
- ❍ How many retailers can a 'normally configured' Windows NT server support?
- ❍ Are retailer programs automatically loaded to the card at the point-of-sale terminal?
- ❍ Can the program run without a server?

Licences and standards

- ❍ TCP/IP data collection
- ❍ 7816-4 cards
- ❍ EMV cards
- ❍ Multos
- ❍ Java card

Patents

- ❍ List of patents held by vendor
- ❍ List of licences to third-party patents
- ❍ Existing or potential infringement

Retailer experience and applications

- ❍ Fastfood
- ❍ Airlines
- ❍ Cinemas
- ❍ Pet stores
- ❍ Supermarkets
- ❍ Retail associations
- ❍ Consumer goods manufacturers
- ❍ Transport authorities

Manage the value chain

A successful deployment requires a well managed value chain. The value chain includes companies involved in various levels of the following activities :

- Card supply, manufacturing, personalization, etc. Companies like Gemplus, Schlumberger and De La Rue manufacture and personalize smart cards. Service companies like First Data Corporation also personalize cards.
- Terminal supply, manufacturing, installation, training, maintenance, etc. Verifone, Ingenico, IVI/Checkmate and Hypercom are some of the best known terminal manufacturers. Thousands of independent sales organizations sell the terminals to retailers, install them, train retailers and offer financing and long-term maintenance and support services.
- Payment transaction acquisition. First Data Corporation is the largest US processor of payment transactions. One of their machines was probably dialled up the last time you used your credit card. Most large banks also offer processing services through subsidiaries like Paymentech, a Bank One subsidiary.
- Card issuing, fulfilment, cardholder relations, etc. This can be done in-house or through numerous service providers. In most cases, it is the card issuer's core activity.
- Loyalty brand management, overall program marketing, merchant recruitment, cardholder recruitment, etc. This is the loyalty brand manager's core responsibility.

Table 9.2 shows the overall value chain. Reading from left to right, cards are used by cardholders, who receive them from card issuers, who receive them from card manufacturers. Finally, the loyalty software provider supplies card manufacturers with the card portion of the loyalty software, thus allowing them to integrate the software during the card personalization phase.

Ongoing support, telephone hot-line assistance and product maintenance become clear as well when viewed through the value chain angle. When a cardholder has a problem, they call the card issuer for support. Card issuers depend on the manufacturer, who in turn depends on support from the loyalty software provider.

Welcome Real-time supplies software, training and ongoing support to card manufacturers, terminal manufacturers, installation and maintenance organizations, payment transaction acquirers and brand managers. Each of these players uses a portion of the loyalty software to enhance their core product offering and in turn markets the software to card issuers, retailers and, ultimately, cardholders.

The loyalty value chain should respect existing agreements and

Table 9.2 Overall value chain

<table>
<tr><th>Object</th><th>User</th><th>Installation, Maintenance, 1st level support</th><th>Equipment/ Infrastructure provider</th><th>Loyalty Software provider</th></tr>
<tr><td>Card</td><td>Cardholder</td><td>Card issuer</td><td>Card manufacturer</td><td rowspan="5">Loyalty software company</td></tr>
<tr><td rowspan="2">Terminal + Retailer PC</td><td rowspan="2">Retailer</td><td>Terminal installation organization</td><td>Terminal manufacturer</td></tr>
<tr><td colspan="2">Payment transaction acquirer</td></tr>
<tr><td>Retailer server (optional)</td><td>Chain</td><td>Terminal installation organization</td><td>Terminal manufacturer</td></tr>
<tr><td>Mileage server (optional)</td><td>Brand manager (or Mileage operator)</td><td colspan="2">Systems integrator</td></tr>
<tr><td>Loyalty brand</td><td colspan="4">Brand manager</td></tr>
</table>

relationships between retailers, card issuers, acquirers and terminal vendors. It is best to integrate loyalty into an existing value chain developed to support the payment infrastructure. Creating a parallel value chain just for loyalty is cumbersome, expensive and lacks the capability of scaling up for a full-size deployment. It is far more efficient to extend the roles of each existing player of the value chain, adding loyalty capabilities at each link in the chain. By leveraging the core competencies of companies already working together, your new smart card program will become far more robust.

InfoAfrica is a South African company that manages the Infinity Card, a nationwide loyalty program targeting tourists. InfoAfrica decided to launch a new generation program using a chip card, this time targeting all inhabitants, not only tourists. It chose XLS, a loyalty software system supplied by Welcome Real-time.

Cards were supplied by Schlumberger, via their local distributor. InfoAfrica simply told Schlumberger what type of loyalty software it was using, and Schlumberger made a proposal for cards that were already compatible with InfoAfrica's software.

InfoAfrica put out a request for proposal for terminals that were also compatible with the selected loyalty software. Ingenico, the world's second largest supplier of payment terminals, happened to be searching for a way to penetrate the South African market. The local distributor had just been established but had not yet made significant progress. Ingenico won the deal and InfoAfrica was able to launch its program.

Retailer training on the terminal was done by Ingenico. Up to this point, there was no need for Welcome Real-time to intervene in the field-level deployment process. In fact, InfoAfrica opted to have transactions processed at Welcome Real-time's headquarters in France so as to minimize the initial investment in people required to manage the system locally. Using a South African Internet access provider, terminals called a local number every night and sent their data collection files to the server's e-mail address (which just happened to be in France).

Once deployment was well under way and the program was functioning properly, it was time to transfer control of the management server over to InfoAfrica personnel. It wasn't until over six months after the pilot had initially been launched that Welcome Real-time personnel actually went to South Africa to install the server and train the client. The efficiency of the value chain meant that each participant brought a specific value-added service to the table without needless overlap of activities and with the least amount of disruption.

A smart card pilot's primary purpose should be to validate and adjust the value proposition. The value chain is a core element of the overall value proposition. As such, your pilot, no matter how small, should be based on a full-scale value chain.

Card issuers are already able to address Schlumberger, Gemplus or any other card manufacturer, and ask for smart cards with the Visa Smart Credit application, for example. In addition, they should be able to ask for cards with the 'Instant Rewards' application, or whatever other loyalty brand they want to use on their cards. Retailers submit similar requests for proposals to terminal manufacturers, specifying that they require the Visa Smart Credit application in the terminal as well as 'Instant Rewards'. Payment processors will also be chosen based on whether or not they support Visa Smart Credit and 'Instant Rewards'.

Launching a pilot without first developing a viable value chain is like going for a test run in a mock-up of a car that doesn't have upholstery, a full dashboard or the high powered engine that is expected for the production model. It might run, but customers won't be very excited about buying one.

It is far more cost effective, and ironically much easier, to run a smart card pilot based on a scaleable, well thought out value chain.

Make sure each value chain participant clearly understands the system's benefits

The brand manager's role is to define how the loyalty system functions for card issuers, retailers and customers. In addition, the brand manager's role is to manage the overall value chain, approving which companies participate and establishing end-to-end pricing. It is necessary to define in very simple and clear terms the main benefits to a company of participating in the value chain.

Why should a card issuer participate? The card issuer will probably be asked to pay a brand licence fee for each card issued with the loyalty brand. The card issuer pays for the right to enter the retail network associated with the brand. In addition, the card issuer may receive a proportion of the transaction fees charged to retailers for loyalty services. Cards with the loyalty brand will tend to be used more often than other cards and cardholders will be easier to sign up and will remain customers longer.

Why should a payment transaction acquirer participate? Acquirers will enjoy increased transactions, in both payment amount as well as quantity, and increased overall fees, due to the additional loyalty fees charged to retailers.

Why should a card or terminal manufacturer participate? Simply because multi-issuer loyalty brands will prove to be the most mainstream use of smart payment cards in many markets. Not participating means the hardware manufacturer will be limited to offering smart cards and terminals to fringe markets.

Why should a terminal sales and maintenance organization participate? Because the added value of being able to provide ongoing support to retailers running electronic punch card programs will prove to be a key area of differentiation. It is much more profitable to sell to a retailer on loyalty functionality than it is to sell on pricing alone.

Prepare for a swift deployment

Once the pilot is made to function properly, deployment can happen very quickly. Your new differentiation strategy based on smart cards is very dependent on the technology in the early phases. But once you begin deployment and momentum builds, your loyalty brand is what will win the battle and guarantee long-term differentiation.

Pilot quickly and deploy even faster.

REFERENCES

❖

1 Sundius, Ann, (1997) *One-on-one interview with the CEO of Visa USA*, MSNBC Business Video.

2 *In the Future ...*, MasterCard International corporate brochure, September 1997.

3 'Chips down for credit card fraud', *BBC News*, 15/03/1999.

4 'A tale of two supermarkets', *BBC News*, 12/04/1999.

5 Waldrop, M. Mitchell, (1996) 'Dee Hock on Management', *Fast Company*, **5**, (10), 79.

6 Collins, James C. and Porras, Jerry I., (1994) *Built to Last: successful habits of visionary companies*, HarperBusiness, pp.142–3.

7 Proton Statistics: *www.proton.be*, and 'VISA; Major Smart Card Players Create Proton World International', *M2 PressWIRE*, 30/07/1998. Carte Bancaire statistics (1997) '5 années de cartes bancaires à puce', *Guide de la Carte*, Analyses & Synthèses.

8 Veverka, Amber, Knight Ridder News Service, 'Smart cards need to be smarter', *San Jose Mercury News*, 09/03/1999.

9 *Relationship Card*, Visa Corporate Relations, corporate brochure, 1995.

10 McKenna, Regis, (1997) *Real-time: preparing for the age of the never satisfied customer*, Harvard Business School Press, p.56.

11 Shapiro, Michael 'Your mileage may vary: Why frequent fliers are often grounded', *The Dallas Morning News*, 22/08/1999, pp.1G.

12 Kelly, Kevin, (1998) *New rules for the new economy: 10 radical strategies for a connected world*, Viking, p.14.

13 'Boots the Chemists: Boots launches loyalty card scheme', *M2 PressWIRE*, 11/08/1997. Also: Cope, Nigel 'Boots shows its hand and joins card wars', *Independent*, 07/08/1997, pp.17.

14 Lin, Wendy, 'Cutting out coupons/Food giants question whether the practice is worth the return', *Newsday*, 03/07/1996, p.B25.

15 'Big Brother at the Supermarket', *Los Angeles Times*, Letter to the Editor, 21/03/1999.
16 Opening quote to *The Great Gatsby* by F. Scott Fitzgerald.
17 Coyne, Kevin P. and Dye, Renee, McKinsey & Co., 'The Competitive Dynamics of Network-Based Businesses', *Harvard Business Review*, 01/01/1998, p.99.
18 Hock, Dee W., (1994) *The One-Horned Cow*, speech given to the Graduate School of Bankcard Management, Norman, Oklahoma.
19 Fulmer, Robert M. and Keys, Bernard, (1998) 'A Conversation with Peter Senge: New Developments in Organizational Learning', Organizational Dynamics, **27** (10).
20 Hock, Dee W., (1994) *Institutions in the Age of Mindcrafting*, speech presented at the Bionomics Annual Conference, San Francisco, California.
21 'OVUM: Software vendors will control booming smart card market predict analysts', *M2 PressWIRE*, 13/05/1998.
22 Gross, Terry, 'How Wal-Mart is Devouring America', *Fresh Air* (National Public Radio), Philadelphia, 19/11/1998.

INDEX

❖

Brain Sell

Tony Buzan and Richard Israel

All selling is a brain-to-brain process, in which the salesperson's brain communicates with the customer's. Recent new discoveries in the fields of psychology, communication, general science, sports and Olympic training techniques, neurophysiology, brain research, sales research and selling techniques have resulted in Brain Sell. In this remarkable book the world's leading expert on harnessing the power of the brain joins forces with a pioneer of modern sales training to show how you can become a high sales producer.

Brain Sell, based on the latest scientific research and the experiences of some of the world's most successful salespeople, explains how to:

- identify which mental skills are currently being used in selling
- apply whole brain selling to any sales situation
- use a multi-sensory format in selling
- develop your sales memory and remember customers' names and faces
- Mind Map and be prepared for the 'sales information age'
- master the mind-body link
- keep focused and retain customer information
- mentally rehearse the sale
- make memorable sales presentations
- develop and use a personal sales commercial.

All of this, together with over 80 skill-building exercises, guarantee a multitude of new ideas in *Brain Sell* for everyone who sells - whatever the type of product or service, and whether you're a beginner or a veteran. Try it!

Gower

Dictionary of Marketing

Wolfgang J Koschnick

The globalisation of marketing and the increasing sophistication of the advertising and marketing techniques that are now used have resulted in a massive growth in the jargon, technical terms and specialized vocabulary used in professional marketing. In addition, developments in computers at the PC and mainframe level now make possible the gathering and analysis of data which measures everything from market share to customer satisfaction. All these developments require at least a working knowledge of a wide spectrum of subjects that range from statistical techniques to the social aspects of target markets. The *Dictionary of Marketing* has been compiled to supply the marketing professional with definitions and explanations of the terms used in these burgeoning fields.

The *Dictionary of Marketing* is designed to be the leading marketing dictionary of the English-speaking world. On some 600 printed pages it lists more than 5,000 terms culled from all areas of marketing. It is the most comprehensive reference book available for marketers and students of marketing and related fields. The book is an attempt to set down an exhaustive range of marketing and marketing-related terms and to provide a definition or, where appropriate, a detailed explanation and description for all entries. In cases where a term or a definition was originated by, or is otherwise closely linked with, a specific person, the name is given in parentheses. A large number of illustrations, charts and tables are included to make some of the more complex entries easier to understand.

In selecting the terms contained in this book, the author has defined marketing widely in order to compile a work of reference that leaves few, if any, questions unanswered.

Gower

Global Sourcebook of Address Data Management

A Guide to Address Formats and Data in 194 Countries

Graham Rhind

Which part of this Asian address is the street? What is this accent and is it correct? Which one of these numbers is the postcode? In which language should I be corresponding? How do I salute this person? In which order should I output this name?

For every individual entered on to a world-wide address database, these questions, and others, need to be answered accurately and correctly. This one-stop reference work covering 194 countries will enable you to have the most accurate international marketing database around - one that makes optimal use of the direct marketing activity generated in your company and ensures that your post is delivered to the correct destination and recipient.

Global Sourcebook of Address Data Management provides, for each country, such information as address and postcode formats, postbox names, salutations, personal name patterns, information about languages, diacritical marks, job titles, casing rules, street types and much more. It will make the management and development of any marketing database more efficient, less expensive and will result in fewer errors and, most importantly, will present the best first impression of the company to its potential customers.

No other book can claim to offer such a comprehensive source of essential information for any manager of an international database, from data quality, direct marketers, market researchers through to telemarketing managers.

Gower

Gower Handbook of Customer Service

Edited by Peter Murley

In a world dominated by look-alike products at similar prices, superior customer service may be the only available route to competitive advantage. This *Gower Handbook* brings together no fewer than 32 professionals in the field, each one a recognized expert on his or her subject. Using examples and case studies from a variety of businesses, they examine the entire range of customer service activities, from policy formulation to telephone technique.

The material is presented in six parts:

- Customer Service in Context
- Measuring, Modelling, Planning
- Marketing Customer Service
- The Cultural Dimension
- The Human Ingredient
- Making the Most of Technology

For anyone concerned with customer satisfaction, whether in the private or the public sector, the *Handbook* is an unrivalled source of information, ideas and practical guidance.

How to Measure Customer Satisfaction

Nigel Hill, John Brierley and Rob MacDougall

The success of your business is underpinned by competitiveness and profitability, both of which are maximized in the long run by doing best what matters most to customers - this book will help you reach that goal.

Written by three leading practitioners, *How to Measure Customer Satisfaction* is a highly practical guide to developing and running an effective customer satisfaction measurement (CSM) programme. To be effective, a CSM programme must first of all produce accurate measures - this book takes readers step-by-step through designing and implementing a CSM survey, highlighting blunders that are commonly made and explaining how to make sure that the measures produced are accurate and credible. It also covers ways of gaining understanding and ownership of the CSM programme throughout the organization, the second key requirement for its long-term success. Finally, the relationship between customer satisfaction and concepts such as loyalty and delight are explored.

If you are committed to the future of your company, the ability to measure what your customers think of you is essential - and so is this book!

Gower